THE GLOBAL AUTOMOTIVE INDUSTRY

Automotive Series

Series Editor: Thomas Kurfess

The Global Automotive Industry	Nieuwenhuis and Wells	September 2015
Vehicle Dynamics	Meywerk	May 2015
Vehicle Gearbox Noise and Vibration: Measurement, Signal Analysis, Signal Processing and Noise Reduction Measures	Tůma	April 2014
Modeling and Control of Engines and Drivelines	Eriksson and Nielsen	April 2014
Modelling, Simulation and Control of Two-Wheeled Vehicles	Tanelli, Corno and Savaresi	March 2014
Advanced Composite Materials for Automotive Applications: Structural Integrity and Crashworthiness	Elmarakbi	December 2013
Guide to Load Analysis for Durability in Vehicle Engineering	Johannesson and Speckert	November 2013

THE GLOBAL AUTOMOTIVE INDUSTRY

Edited by

Paul Nieuwenhuis
Cardiff University, UK

Peter Wells
Cardiff University, UK

WILEY

This edition first published 2015

Registered Office
John Wiley & Sons, Ltd, The Atrium, Southern Gate, Chichester, West Sussex, PO19 8SQ, United Kingdom

For details of our global editorial offices, for customer services and for information about how to apply for permission to reuse the copyright material in this book please see our website at www.wiley.com.

Library of Congress Cataloging-in-Publication Data applied for

ISBN: 9781118802397

A catalogue record for this book is available from the British Library.

Set in 10/12pt Times by SPi Global, Pondicherry, India

Printed in the UK

Contents

Notes on Contributors

Editors' Profiles

Paul Nieuwenhuis
Centre for Automotive Industry Research and Electric Vehicle Centre of Excellence
Cardiff Business School, Cardiff University
Cardiff, Wales, UK

Dr Paul Nieuwenhuis is a senior lecturer at Cardiff University. He joined the Centre for Automotive Industry Research (CAIR) at Cardiff University in 1991, and he became one of its two directors in 2006. He was a founder member of the ESRC Centre for Business Relationships, Accountability, Sustainability and Society (BRASS) and is an associate of the Sustainable Places Research Institute. His main interests are historic and environmental, and his publications have been in these areas, for example *The Green Car Guide* (1992) and *Sustainable Automobility* (2014). He also contributed to the *Beaulieu Encyclopaedia of the Automobile* (2000), which won a Cugnot Award from the Society of Automotive Historians. Dr Nieuwenhuis has produced around 300 publications ranging from books and academic papers, to conference papers for both academic and business audiences and journalistic pieces. Dr Nieuwenhuis is a member of the Guild of Motoring Writers.

Peter Wells
Centre for Automotive Industry Research
Cardiff Business School, Cardiff University
Cardiff, Wales, UK

Peter Wells is a professor of sustainable business models at Cardiff Business School, where his work has ranged across spatial industrial development, economics, organizational theory, industrial ecology, technological change, transition theory, business models and sustainability – all through an applied focus on the global automotive industry. Professor Wells has over 550 publications reaching academia, industry, policy and stakeholder audiences through traditional papers and books, Internet publications and, more recently, webinars.

Contributors' Profiles

Katsuki Aoki
School of Business Administration
Meiji University
Tokyo, Japan

Dr Katsuki Aoki is an associate professor in the School of Business at Meiji University in Japan. He received his Ph.D. in business administration from Meiji University in 1999. His main research interests include (i) international comparative studies on the implementation of kaizen activities, (ii) the benefits and limitations of the keiretsu system (OEM–supplier relationships) in the automotive industry and (iii) mass customization and order fulfilment systems in the automotive industry. His paper entitled 'Transferring Japanese kaizen activities to overseas plants in China' was selected as one of the most prominent papers at the Emerald Literati Network Awards for Excellence 2009.

Liana M. Cipcigan
Electric Vehicle Centre of Excellence
School of Engineering
Cardiff University
Cardiff, Wales, UK

Dr Liana M. Cipcigan is a senior lecturer in the School of Engineering at Cardiff University, leading the research of integration and control of EVs in electricity and transportation networks. She is a member of the Electric Vehicle Centre of Excellence (EVCE). Her current research activities are focused on smart grids, distributed generation and EV integration and control. She has significant research experience in EU projects: FP6 More Microgrids, FP7 Mobile Energy Resources in Grids of Electricity (MERGE) and ERDF ENEVATE. She is the principal investigator (PI) of the EPSRC projects 'Smart Management of Electric Vehicles' and 'Electric Vehicle Value Chain – Bridging the Gaps'. She is participating in the UKERC project 'Smart Grids Scenarios for UK' and TSB project 'Agent-Based Controllers for Electric Vehicles and Micro-generators'. She is a member of CEN-CLC eMobility working group on Smart Charging and IEEE P2030.1 working group on standards for EVs and related infrastructure.

Huw Davies
Electric Vehicle Centre of Excellence
School of Engineering
Cardiff University
Cardiff, Wales, UK

Dr Huw Davies is a lecturer in mechanical engineering at Cardiff University School of Engineering. His background is in vehicle engineering and the development of vehicle safety standards. He has advised the UK Department for Transport, the European Commission and the Automotive Industry. At Cardiff University, he has developed the transport research theme. Safety, mobility and emissions are at its core. The universal goals are zero collisions, zero congestion and zero emissions. Dr Davies leads Cardiff University's Electric Vehicle Centre of Excellence.

Ceri Donovan
Electric Vehicle Centre of Excellence
School of Engineering
Cardiff University
Cardiff, Wales, UK

Ceri Donovan is a research assistant at Cardiff University School of Engineering and a member of Cardiff University's Electric Vehicle Centre of Excellence. She spent most of her time in recent years working on ENEVATE, which aimed to accelerate the uptake of e-mobility in northwest Europe. This includes conducting research on all aspects of electric vehicles, including market drivers and mobility concepts, vehicle safety and regulations. She previously worked in defence research, primarily focusing on how to integrate new technology onto existing platforms from the procurement as well as the practical perspectives. She has an M.Sc. in biometry from the University of Reading.

Patrick Galvin
Innovation Policy Lab
Munk School of Global Affairs
University of Toronto
Toronto, Ontario, Canada

Patrick Galvin is a postdoctoral research fellow with the Innovation Policy Lab at the Munk School of Global Affairs, where he is part of a research team investigating the dynamics of developing a new automotive policy for the Canadian automotive industry for the twenty-first century. He is also working on the SSHRC-funded Creating Digital Opportunity project. Prior to his current appointment and after having obtained both his B.A. (honours) and M.A. degrees in political science, Patrick spent several years working in the housing policy field with a number of consulting firms. Patrick enrolled in the Ph.D. programme in politics at the University of Exeter in England. He completed his dissertation in November 2012, and he was formally awarded his Ph.D. degree in July 2012. Patrick's Ph.D. dissertation built on his training in public policy and political economy by examining how local government in the city of Toronto develops its cluster-related innovation policy. It focuses on two empirical case studies to see how the city develops its cluster policy for two industrial sectors: the aerospace sector and the fashion sector.

Elena Goracinova
Innovation Policy Lab
Munk School of Global Affairs
University of Toronto
Toronto, Ontario, Canada

Elena Goracinova is a Ph.D. candidate in political science working at the University of Toronto's Munk School of Global Affairs Innovation Policy Lab with Professor David Wolfe. She received her master's degree in geography from the University of Toronto with a thesis on manufacturing in developed economies. She is interested in studying the role of the state in economic policymaking. Her current work focuses on the scope and effectiveness of advanced manufacturing policies in Canada.

John Holmes
Professor Emeritus
Department of Geography
Queen's University
Kingston, Ontario, Canada

John Holmes is a professor of geography at Queen's University in Kingston, Ontario, Canada. He received his B.Sc. (honours) and M.A. (social science) degrees from the University of Sheffield and his Ph.D. from Ohio State University. He is also affiliated with the graduate industrial relations programme in the Queen's School of Policy Studies. His research focuses on geographical aspects of the political economy of contemporary economic and social change and, in particular, on the contemporary restructuring and reorganization of production and work in North America. Empirical research and writing have focused primarily on the automobile industry. For a list of selected publications, see http://geog.queensu.ca/faculty/holmes.asp.

Seunghwan Ku
Faculty of Business Administration
Kyoto Sangyo University
Kyoto, Japan

Seunghwan Ku is a professor of technology of management of the Faculty of Business Administration, Kyoto Sangyo University. He received his Ph.D. (management of technology) from the University of Tokyo, Japan. His dissertation entitled 'The Dynamism of Product Architecture: Modularization, Knowledge Integration, Interfirm Linkage' was published by Mineruba Shobo, a major publisher in Japan. His recent publications include 'Economic Analysis of ICT Innovation' (2011) with M. Fujiwara and 'Ship Development and Shipbuilding Industry: The Constraint of Large Artifact Development and the Uncertainty of Business Model' with H. Kato in T. Fujimoto, ed., 'Coping with Complex Artifacts' (2013). His current research interests are (i) product development, (ii) logistics and inter-firm linkages, (iii) product strategy and innovation and (iv) supplier system in Korean and Japanese industries, the auto sector in particular.

Maneesh Kumar
Centre for Automotive Industry Research
Cardiff Business School, Cardiff University
Cardiff, Wales, UK

Dr Maneesh Kumar is a senior lecturer at Cardiff Business School, Cardiff University, UK. His research interests are primarily in the area of quality management, Lean Six Sigma (LSS) and service recovery within manufacturing, service and public sector organizations. His recent research within Indian automotive industry focuses on understanding the best-in-class practices in diffusing Lean/Kaizen practice at supply network level and how automotive giants (OEMs) support and develop their suppliers (at Tier 1 and Tier 2 level). His research outputs include an edited book, five edited conference proceedings, three book chapters and over 80 peer-reviewed journal publications and conference papers. He has been involved in delivering LSS training up to black-belt level and delivered several workshops on LSS application in several blue chip companies. He is also a regular speaker at international conferences and seminars on LSS and process excellence.

Xiao Lin
Centre for Automotive Industry Research
Logistics and Operations Management Section
Cardiff Business School, Cardiff University
Cardiff, Wales, UK

Xiao Lin is a Ph.D. student in the logistics and operations management section at Cardiff Business School, Cardiff University. Her research interests include sustainability, transition theory, business models, etc. Her Ph.D. research project is specifically focusing on the study of the future prospect for e-bikes in sustainable mobility in China, through various quantitative and qualitative analysis methods. Xiao Lin is enthusiastic in contributing to the collaboration of the sustainable community. She is a member of the organizing committee of the 'The 2014 Global Research Forum, Sustainable Production and Consumption Conference'.

Daniel Newman
Electric Vehicle Centre of Excellence and Sustainable Places Research Institute
Cardiff University
Cardiff, Wales, UK

Dr Daniel Newman is a research assistant at Cardiff Business School, working within the Cardiff University Electric Vehicle Centre of Excellence. His activities are primarily focused on the ENEVATE project, conducting social research investigating the potential for increasing electric vehicle uptake in northwest Europe. He was previously a research assistant at the University of Bristol Law School, looking at environmental law and health and safety regulations. Prior to this, he completed an ESRC-funded Ph.D. in the University of Bristol Law School, investigating the state of access to justice, which was published as a book by Hart in 2012.

Ben Waller
Senior Researcher, ICDP
International Car Distribution Programme (ICDP)
Cardiff, Wales, UK

Ben is a senior researcher at the International Car Distribution Programme (ICDP) and based in the United Kingdom. ICDP is an international research-based organization focused on automotive distribution, including the supply and retailing of new and used vehicles, aftersales, network structures and operations. Through research activities, data services, education, events and consulting, ICDP works with vehicle makers, dealers, suppliers and related organizations to improve the quality and effectiveness of the distribution model. Ben has presented to a range of international audiences, from academic conferences to senior executives, and also authors articles for business press and delivers executive teaching.

Lorraine Whitmarsh
School of Psychology
Cardiff University
Cardiff, Wales, UK

Dr Lorraine Whitmarsh is an environmental psychologist, specializing in perceptions and behaviour in relation to climate change, energy and transport. She is a senior lecturer in the School of Psychology at Cardiff University, UK. Lorraine is also a partner coordinator for the Tyndall Centre for Climate Change Research and a research associate of Cardiff University's Centre for Business Relationships, Accountability, Sustainability and Society (BRASS) and the Sustainable Places Research Institute. She sits on the Climate Change Commission for Wales, advising Welsh government on transport and behaviour change.

David A. Wolfe
Professor of Political Science and
Co-director, Innovation Policy Lab
Munk School of Global Affairs, University of Toronto
Toronto, Ontario, Canada

David A. Wolfe is a professor of political science at UTM and a co-director of the Innovation Policy Lab at the Munk School of Global Affairs in Toronto. From 2009 to 2014, he was the Royal Bank Chair in Public and Economic Policy at the University of Toronto. He is currently leading a major partnership grant on *Creating Digital Opportunity: Canada's ICT Industry in Global Perspective* funded by the Social Sciences and Humanities Research Council of Canada. He is also participating in the project on manufacturing policy and the Canadian automotive sector based in the Automotive Policy Research Centre at McMaster University.

Dimitrios Xenias
School of Psychology
Cardiff University
Cardiff, Wales, UK

Dr Dimitrios Xenias is a social and environmental psychologist, focusing on sustainable travel and associated behaviours, such as mode choice; on domestic energy-related behaviours; and on perceptions of climate change and its communication. He is a research associate in the School of Psychology at Cardiff University, UK. Dimitrios is also an associate of the Tyndall Centre for Climate Change Research and a research associate of Cardiff University's Centre for Business Relationships, Accountability, Sustainability and Society (BRASS) and the Sustainable Places Research Institute.

Series Preface

The automotive industry is one of the largest manufacturing sectors in the global community. Not only does it generate significant economic benefits to the world's economy, but the automobile is highly linked to a wide variety of international concerns such as energy consumption, emissions, trade and safety. The primary objective of the *Automotive Series* is to publish practical and topical books for researchers and practitioners in the industry and postgraduates/advanced undergraduates in automotive engineering.

The series addresses new and emerging technologies in automotive engineering supporting the development of more fuel-efficient, safer and more environmentally friendly vehicles. It covers a wide range of topics, including design, manufacture and operation, and the intention is to provide a source of relevant information that will be of interest and benefit to people working in the field of automotive engineering.

This book, *The Global Automotive Industry*, presents a broad spectrum view of the automotive sector from the technology, industry and policy perspectives. In the early chapters of the book, an understanding of the industry from a social, technical and economic perspective is presented. This is followed by a discussion of the fabric of next-generation vehicle systems as it is woven into a global society. This discussion is expanded to include the economic impact of the automotive industrial sector on nations' economies from production operations/revenue to job growth. This text goes beyond regional thinking with respect to automotive production and discusses the historic expansion of automotive manufacturing giants in Korea and the new emerging markets in China and potentially India.

The text finishes with an excellent overview of emerging technologies and economies that will be pivotal to the automotive sector. These discussions are then presented in the light of new business models in a global marketplace, and the impact of regulation on the automobile as well as its manufacture. These topics are well integrated into the earlier topics presented in the book and make it a unique and valuable crosscutting source of information for the reader.

The Global Automotive Industry integrates these topics in a unique and thought-provoking manner that provides significant insight into global automotive production, regulation and economic impact on society. It is a forward-thinking text that considers a wide variety of topics that will drive this major industry and the global economy for the foreseeable future. It is written by highly recognized experts in the field and is a welcome addition to the *Automotive Series*.

Thomas Kurfess
July 2015

Foreword

The global automotive industry is a complex and demanding topic for any book. It is clear following my 30-year career within the automotive industry that only a comparatively small number of individuals have a comprehensive understanding of a vehicle's full construction. Even after decades of professional involvement, it can be difficult for an individual to have a full appreciation of the subject and its international infrastructure and landscape.

From pre-production to the factory floor and the boardroom, knowledge sharing is incredibly important to the future of the automotive engineering profession. With future mobility solutions being vital to modern society and with the needs of each region differing depending on capabilities and production facilities, communication amongst the world's engineers provides a route to approaching and accommodating the global social, political and economic considerations that the future will demand.

FISITA has successfully coordinated global communications amongst its network of over 200 000 professional engineers from member societies in 37 countries for over 60 years. This demonstrates the point through many generations that sharing of knowledge and ideas amongst an industry as large as automotive, including engineers, academics and executives, is vital as it provides us all with the tools to contribute to a better future.

As part of this journey, global collaboration within the industry is also necessary in order to deliver a clear message to wider society: that the automotive engineering profession is well placed to face the challenges of today and tomorrow.

Paul Nieuwenhuis and Peter Wells have provided us with a very good starting point. The authors are able to present the wide scope of the industry in a clear way whilst contributing a wealth of knowledge about the global automotive industry.

Chris Mason
CEO, FISITA

1

Introduction and Overview

Paul Nieuwenhuis[1] and Peter Wells[2]
[1] *Centre for Automotive Industry Research and Electric Vehicle Centre of Excellence, Cardiff Business School, Cardiff University, Cardiff, Wales, UK*
[2] *Centre for Automotive Industry Research, Cardiff Business School, Cardiff University, Cardiff, Wales, UK*

1.1 Introduction

It is an impossible task to do full justice, in one book, to the global automotive industry with the pervasive impact, economic significance and cultural status that it brings. As editors, our task is as much one of selective omission as one of collation, to create an account of the business of making, selling and using cars that is both representative and yet respectful of the diversity we know to be a feature of the industry. Moreover, this is an industry in the midst of reinventing itself. It had seemed to some both inside and outside the industry, around the turn of the millennium, that this was a 'sunset' industry characterized by over-capacity, plant closures, job losses, declining profitability and a product that seemed unsuited to meeting the environmental and social challenges arising from mass car ownership and use. The combination of profound environmental pressures and endemic economic distress appeared to call into question the pre-eminent position enjoyed by the car in providing for personal mobility and the viability of the industry behind this most contradictory of products. These concerns came to a peak with the arrival of the global economic crisis around 2007–2008, which resulted in new car sales falling steeply in the established markets and investment freezing while the entire financial community held its collective breath to see whether the world would plunge into an economic abyss. Those looking closely at urban developments and the changing cultural priorities of young people started to whisper about the concept of 'peak car' in which the high tide of automobility had been reached in the saturated markets of the European Union, North America, Japan and Korea.

Yet, only a few years later, the industry has reemerged, burnished by the embrace of new technologies and with resurgent sales in the previously moribund West being more than matched by the frenetic growth of sales in China and elsewhere. Regulatory interventions

The Global Automotive Industry, First Edition. Edited by Paul Nieuwenhuis and Peter Wells.

such as the European Union fleet average carbon emissions targets that were once seen as burdensome and punitive have arguably helped stimulate the technological renaissance of the industry alongside the realization that there might be other routes to profitability. Later chapters in the book are more concerned with this recent period of transformation and the possible futures it portends: Chapter 16, for example, is about the impact of the electric vehicle, while Chapter 18 links technology change with possible innovations in business models forced on the industry by such technological changes. Chapter 19 is more speculative, but offers up some signposts for the future of automobility and the automotive industry.

This is not to say that the process of change has been without stress or consequence. Restructuring of the industry has continued apace: the merger of Chrysler and Fiat came as a consequence of several years of corporate turmoil while the rescue of PSA by the French government and by the Chinese company Dongfeng costs the Peugeot family their control over this long established business – the oldest in the industry. The great expectations held for battery electric vehicles remain as yet largely unrealized, despite the highly visible and expensive efforts of companies like Nissan, Renault and BMW alongside the publicity-garnering machine that is Tesla. In the meantime, high-profile failures such as the battery swap business Better Place serve as stark reminders that attempts to reinvent this industry, especially by new entrants, are fraught with hazard. At the same time, markets in the European Union have not really recovered to pre-crisis levels and perhaps never will.

So, despite an apparent 'business-as-usual' attitude permeating the automotive industry at the moment, in reality, it is subject to potentially radical change. Only a few years ago, we saw two of the world's largest car manufacturers, General Motors and Chrysler, on the brink of collapse. Having decided both were 'too big to fail', they had to be bailed out by the American tax-payer. Ford managed without direct government support, but it is nevertheless quite clear that the collapse of the supply base that would have resulted from a collapse of both their key domestic competitors, GM and Chrysler, would also have dragged down Ford in due course. Within a matter of months, the world's most prominent car industry and the oldest mass production car industry could have disappeared; only intervention from the public sector saved this key component of the private sector, underlining that, apart from anything else, the close synergies and intertwined relationships continue to exist between the private and public sectors in developed countries (Mazzucato, 2013). This highlights not only the weaknesses inherent in the close links between the financial and car systems but also indicates that the current mass production system, whether seen as Fordist or Buddist (see below), may well be due for a rethink in several respects.

In Europe, the apparent success of Germany, more recently the UK automotive industry, is easily misunderstood, leading perhaps to overconfidence, but do note that the secret of success in Europe is largely confined to the more upmarket segments, with the Germans the primary beneficiaries, and that this success is very reliant on demand from Asia, particularly China; it has therefore made Europe – and also the United States, particularly GM – very dependent on that far from stable market, as also outlined in more detail in Chapter 10. At the same time, pressures to make the car more environmentally compatible, as we see it, have not only led to cars with much less harmful tailpipe emissions, but the more recent pressure to reduce the carbon footprint of cars, particularly focused on their emissions of the greenhouse gas carbon dioxide, has the potential to radically reshape the industry and its products. These regulatory pressures are reviewed in Chapter 14, although the potential impact of this agenda is not yet fully understood, but it has already led to a new wave of alternative powertrain vehicles in the market, ranging from early petrol-electric hybrids – including the now iconic Toyota Prius – via so-called range-extended and plug-in hybrids such as the GM Volt and Ampera, to full battery-electric vehicles such as the Nissan Leaf, Renault Zoe and Tesla Model S. Such developments, as well as the introduction of the even more radical BMW i3, which not only introduces a full battery-electric powertrain, but also houses this within a radical departure from the tried and tested 'Buddist' (see Chapter 5) all-steel body in using a carbon fibre body on an aluminium chassis, are clear signs that business-as-usual is steadily being undermined even from within the established industry.

Despite the size of the industry, car making is in reality a very precarious business. In its present form, it lacks resilience; it is not sustainable. This is true not just in terms of environmental sustainability, where it clearly is not measuring up, but even in terms of basic, economic sustainability. Yet, this industry is still crucially important, representing a significant part of the economy in many industrialized countries, as well as many newly industrializing economies, notably China. In addition, the car-based transport system, or 'regime' (Geels et al., 2012), which extends well beyond the industry that makes cars to the way in which automobility has become integrated in our societies and cultures, has become so embedded that its removal would lead to widespread social and economic crisis.

In some regards, therefore, this book may be a testimony to a dying era as much as a hymnal to a new one. Thus far the automotive industry has managed to contain the pressures for change, both economic and environmental, within the broad ambit of the long-standing business model of the mainstream vehicle manufacturers. In parallel, those purchasing and using cars in the traditional manner have remained by far the majority compared with more innovative solutions such as car clubs, city car sharing schemes and other alternatives that might presage the end of established automobility cultures. The question is whether the collective and cumulative impact of all the new developments sweeping through the industry will be sufficient to herald the arrival of a new 'automotive ecosystem' as some have termed it, or 'regime', according to others, and one in which the dominance of the established vehicle manufacturers and their entrenched supply chains is challenged by a new order of communications and mobility providers. Some caution on these issues is urged here. The automotive industry is not immune to hype cycles or the fevered imaginations of those professional change-mongers with a vested interest in fermenting and emphasizing the new over the continuous.

1.2 Continuity and Change

The book as a whole therefore represents an attempt to capture continuity and change in the global automotive industry; this theme is brought to the fore in the first main chapter of the book, Chapter 2, as change is very much a feature of the current phase of the automotive system. Change of course needs to be seen in historical perspective and as such, many of the chapters offer up a historical narrative that builds a temporal dynamic into the account of the subject under consideration. There is some need, in seeking to understand the structures and practices of the present, to delve into the past. If nothing else, path dependency theory tells us that many decisions are nonlinear, irreversible steps that result in distinct historical outcomes bounded by time and place. As we have sought to delineate elsewhere, the cluster of innovations that created the template for the mass production automotive industry occurred largely in the United States in the early years of the twentieth century: the Ford moving assembly line with standardized and inter-changeable parts; the Budd all-steel body that could be stamped, welded and painted and the Sloan multi-brand company with marketing innovations such as consumer credit and annual model changes; these developments are outlined in Chapter 5. Despite many years of incremental change, the basic architecture of the industry remains in place. That such durability is evident speaks volumes for the scope for productivity improvement within this basic architecture over many decades and the efficacy of the system as a whole in terms of matching the imperatives of production economics to the demands of the market. For this reason, key early chapters in the book, for example Chapters 3, 4, 5 and 6, are about the production system and the market for cars. There is no subscription here to the simple neoclassical economics assumption that 'consumers' demand and 'producers' provide. Rather, there is a continuing dynamic tension between supply and demand sides that is never entirely resolved.

1.3 Overview

The chapters in this book are thus essentially an attempt to take stock of the industry in its present state of a well-established industry on the verge of potentially radical change, tracking some of its history – trying to answer the question of how did we get here – assessing key aspects of its current state, while also plotting some possible futures for the industry and its products. This book does not seek to present chapters that are held together by a common theoretical thread. Such works can be tremendously informative and insightful (see Geels et al., 2012), but of course tend to place primary emphasis on the question of theoretical coherence. Rather, the approach here has been more agnostic in theoretical terms and has been to seek insights and individual views by those who are established experts in their fields and who are known for their knowledge of the automotive industry (or a specific aspect of it) alongside their particular theoretical or methodological frameworks. Each chapter is thus a cameo, able to stand alone as a concise insight into one dimension of this multi-faceted industry but also serves as a piece in the jigsaw that describes and explains how this industry really works.

There are some gaps of course. It is not possible to give consideration to all the countries producing cars and the unique contexts within which they operate. The salience of the 'varieties of capitalism' school of thought (see Section 2.3) is acknowledged here and some comparative chapters are pertinent to this school: notably those on labour–management relations in North America and Japan (Chapters 7 and 8, respectively). While the question of labour relations in Japan has remained largely immune to external influence (notwithstanding the substantial impact of Renault on operations at Nissan and the rather lesser impact of Ford on Mazda), the story in North America has a strong thread in which 'Japanization' has been highly significant. Chapters 9, 10 and 11 cover the broad sweep of the automotive industry in Asia, highlighting South Korea, China and India respectively – both Korea and India have become comparatively neglected in view of the immense attention directed at China. Interestingly, both these countries are more than passive recipients of automobility and can claim to be home to significant companies within the industry of the modern era, while also representing unique markets.

An apparent gap in the account includes a chapter that explicitly deals with the global supply chain of components and raw materials to the vehicle manufacturers – where typically up to 85% of the ex-works value of a new car is accounted for by bought in components and materials. The theme of supply chain management is an important facet of the industry but is not given a distinct treatment in this book, although aspects of it are covered in a number of chapters. Indeed, a more 'functional' approach to the industry could include themes such as marketing and branding, inbound and outbound logistics, vehicle finance, production management, R&D, government affairs, human resource management, leadership, trade and currency risk management, reverse logistics, service and maintenance, end-of-life vehicle management and much more. Inevitably, aspects of these functional typologies do in any case appear to a greater or lesser extent in the chapters, but the book is deliberately not designed around them.

On the car supply side, the automotive manufacturing industry itself, the debate has long been dominated by issues surrounding the organization of labour and this is reflected in Chapters 7 and 8 in particular, which highlight the traditional mass production, or 'Fordist', or 'American System'-based approach, while contrasting this with the different Japanese approach, as developed under the Toyota Production System, now also known as 'lean' production. Both chapters also highlight how in recent years the two approaches have increasingly influenced each other, a theme also touched upon in Chapter 5. Chapters 6 and 9 take a more mixed approach, with the former reporting on and analysing recent trends in mass manufacturing in various locations around the world and the implication of these on the organization of labour. Chapter 9, while still involving discussion of the organization of labour in a mass production system, also proposes the role of other factors in the rapid growth of the Korean car industry, something also reflected in Chapter 11 on India. The other feature of the supply-side story is the role of technology; it was technology after all that enabled the development of mass production, which then prompted a rethink on the organization of labour used to interface with that technology. This is the dominant theme of Chapter 5, which recaps previous work on the technological history of mass production, as pioneered by Edward Budd and links this with recent advances in 'lean thinking'.

The demand-side story is interwoven with the above chapters, as in truth supply and demand are closely linked. This side of the automobility system is first covered in Chapter 3, which deals with markets, while Chapter 4 analyses the crucial psychology of automobility in all its different aspects – how do people make decisions about their transport needs? Chapter 12, then, covers the interface between supply and demand more specifically: the crucial retail and distribution system, which is tightly linked in with the way cars are mass produced today, being tasked with resolving many of the issues that arise from the requirement, outlined above, to reconcile the needs of the mass production system for standardization, with the needs and wants of the market for increasing differentiation. The three chapters that follow then consider wider aspects of the automobility system. Thus, Chapter 13 provides an account of the various negative impacts of the industry, particularly in the areas of occupant safety and toxic emissions; Chapter 14 examines how such impacts have been ameliorated through regulation, while Chapter 15 considers the bigger picture of the role of the industry in the world, the ongoing globalization process and the resulting impacts on economies both global and local.

The final four chapters, then, consider the future, which in product terms is likely to witness an increasing adoption of electric powertrain vehicles of various types, ranging from simple petrol-electric hybrids and diesel-electric hybrids, via plug-in and range-extended hybrids to battery-electric and fuel cell cars. These are discussed in Chapter 16, while Chapter 17 reviews various possible alternatives to the car and the current car system. Chapter 18 then discusses the need for new technology and use configurations to be delivered through new business models that may be rather different from those that both industry and end user have become familiar with over the past 100 years or so. The final chapter then considers the state of the industry and its future prospects in the light of what has been covered in the other chapters.

The story of the car industry and the automobility system it has allowed to be created is a compelling one. However, many fundamental aspects of both are likely to come under increasing pressure in the very near future. An overview of the industry and the socio-technical regime that underpins it is therefore timely. This book contains insights by experienced and in many cases world-renowned experts in the industry in its various aspects, thereby presenting a real state of the art of the industry today, its explanation in the past and prospects for its future.

References

Geels, F., Kemp, R., Dudley, G. and Lyons, G. (2012) *Automobility in Transition? A Socio-Technical Analysis of Sustainable Transport*, Vol. **2** Routledge Studies in Sustainability Transition, New York: Routledge.

Mazzucato, M. (2013), *The Entrepreneurial State: Debunking Public vs. Private Sector Myths*, London: Anthem Press.

2

Understanding Change and Difference in the Global Automotive Industry

Peter Wells
Centre for Automotive Industry Research, Cardiff Business School, Cardiff University, Cardiff, Wales, UK

2.1 Introduction

The global automotive industry is one in which change does appear to be endemic, and much has been made of the processes of globalization whereby a single coherent industry may emerge. Yet, it is apparent that understanding the degree and pace of change in the industry, and of whether or not there is convergence and integration at a global level, requires more than the reductionism of globalization. The structural volatility of the sector alone is sufficient cause to think that a simplistic consolidation into an ever-diminishing clutch of multi-brand, multi-market companies is unlikely to materialize in the near future whether at vehicle manufacturer or supplier level. The additional complicating interventions of the state at multiple levels further acts to erode any suggestion of 'pure' economic forces shaping the industry. Then, the insertion of cars into everyday life results in a sort of hybridization of cultures of automobility, with differing flavours of such automobility emerging in different places despite the apparent homogeneity and ubiquity of the car.

This chapter therefore seeks to draw on three distinct literatures and perspectives with regard to the theoretical understanding of change and non-change in industrial structures. It is recognized that of course there are other understandings or insights from other theoretical traditions, but the perspectives offered by *socio-technical transitions*, *varieties of capitalism* and *global value chains* respectively have been both influential and heavily involved in empirical study of the automotive industry. Each of these theoretical traditions seeks, in its own way, to step outside the narrowly economic dynamics of the industry and look for wider

The Global Automotive Industry, First Edition. Edited by Paul Nieuwenhuis and Peter Wells.

societal structures for the explanation of change and stability. Understanding the relationship between continuity and change is a continuous challenge for those engaged in business and economic research. There is a bias towards change because novelty and 'news' sell – not just newspapers but also academic journals. There is also a bias towards 'success' and seeking to explain what went right (particularly in the mainstream business literature) so that others might learn from and emulate those successes.

As a consequence, there is less effort invested in seeking to understand inertia, failure or active resistance to change and perhaps also a tendency to focus on the big companies, the emblematic technologies and the major government policy initiatives in a manner that can easily miss the collective and cumulative impact of minor developments. Underpinning this understandable tendency to focus on the apparently most important companies, technologies and policies is a largely unspoken sense of the difference between incremental and radical change. Sometimes, clear discontinuities are evident: a significantly different working regime may be introduced, for example, or a significantly and qualitatively different technology may penetrate the market. Often, profound change is more obvious in hindsight than in the present, but simultaneously our expectations over the future rate and pace of change are often rather optimistic given historical experience.

Concepts of change and inertia are inevitably underpinned by a (often implicit) notion of time. A change in a system or a phenomenon means that it is different in one time point compared with another. Words like 'radical' or 'revolutionary' change imply a rapid change over a short time, compared with 'incremental' or 'marginal', which imply a rather modest change spread over a longer time period. This much is clear. What is rather less clear is how this is treated within different theoretical traditions and discourses, and what metrics are available to allow us to measure change over time in a manner that then allows judgements to be made on whether such changes constitute radical or incremental change. In the socio-technical transitions literature, the early examples of regime shift (i.e. a transition from one socio-technical regime to another) unfold over a period of 80 years or more (Geels, 2002). In contrast, global value chain theory really engages in understanding shifts in the global economy over the recent past (say the last 20 years) and the near future. These differences in temporal framing may be contributory to isolating one line of academic inquiry from another, or make the task of integrating different theoretical perspectives that bit more challenging.

Academia also tends to layer into the discussion certain biases, one of which is the quest for generalization as a means of attaining a higher level of truth. Hence, the more we are able to say that the phenomenon that our theory seeks to define and our data seek to illuminate is generalizable, the more powerful is the truth that we have to offer. In the social sciences, such a quest is often problematic because explanations, particularly narratives as are found elsewhere in this book, tend to be grounded in context and specificities. Again, the three theoretical approaches outlined in this chapter all in their various ways seek to deploy a theoretical framework that transcends time and place, but that in application are verified by and in specific concrete conditions.

It is pertinent at this point also to consider the question of new entrants and incumbents, as these are often considered to have different capacities for change. Indeed, perhaps reflecting a certain ideological bias towards entrepreneurial actions, there is often a distinct inclination to consider that new entrants are the vectors of change, while incumbents are passive or active in resisting change. Yet in highly complex, rapidly evolving and inherently unstable socio-economic structures and practices, this neat distinction is rather difficult to sustain. It could be argued that incumbents are best placed to be agents of change: they have most to lose from the destabilization of the current order, and they have the resources (knowledge and skills; physical capital and infrastructure; financial resources; supply chains, contacts and networks; brands and customer bases) to make change happen. Alternatively, new entrants may manage a precarious existence on the margins of an industry, but fail to penetrate into anything approaching significance, and if by chance they do become important then they may be subsumed by the dominant incumbents. Moreover, the incumbent/new entrant distinction rather ignores other potentially important vectors of change, in particular the national or sub-national state. The state may act in many distinct ways as a catalyst and participant of change: from funding R&D through to being a significant 'stepping stone' market for innovative technologies. Indeed, Mazzucato (2013) argues quite convincingly that all major technological innovations have seen public sector involvement in their early stages, from the worldwide web, through GPS to space exploration. Only after such technologies have become sufficiently established can viable markets be recognized and can private entrants justify the now reduced risk of entering such novel markets (Mazzucato, 2013). However, beyond the formal structures of the state are the more nebulous but still highly important impacts of social practices, behaviours, attitudes and norms that constitute much of what makes one place different to another, and what social forces act to constrain or enable change to happen.

The chapter proceeds in the following manner. Each of the three main perspectives of change under consideration here is considered in isolation and in turn: from socio-technical transitions, through varieties of capitalism, to global value chains. Thereafter, a more speculative section seeks to draw some of the possible explanatory merit to be derived from bringing insights from each individual tradition to bear on the others. While this final section is by no means comprehensive, it does at least suggest a research agenda that finds an empirical focus in uncovering the unfolding of technological, social and economic changes as they play out across geographic space. The process of theory building in the social sciences often builds new insights from the hybridization of previously distinct schools of thought in a multi-disciplinary or cross-disciplinary manner, and this chapter therefore marks a further attempt to continue this tradition.

2.2 Socio-Technical Transitions

Socio-technical transitions are conceived of as long-term structural shifts in the economy and society arising from the permeation of key technologies into widespread use. A dynamically stable set of inter-relationships may then develop linking technologies with economic structures (such as markets and international trade relationships), organizations (such as companies, Trades Unions and NGOs), government (such as in the form of regulations and laws, and participation in the market) and social practices (including behaviours and beliefs). The resulting ensemble is often termed a 'regime', which is not only enduring, but effectively acts to secure the reproduction of the conditions for its own existence.

In this theoretical framework then there are two main areas of interest: the significance of path dependency in structuring nonlinear change and indeed resistance to change, and the scope for shifts in the existing regime. Regime stability may be undermined from one of three main sources in this theoretical concept: meta-pressures at what (in the so-called multi-level perspective) is called the 'landscape' level; changes from within the regime itself; and what are termed 'niche' developments that may grow and develop to an extent that they displace the existing regime (Geels, 2002). The landscape consists of external framing conditions in the form of resource availability, infrastructure, broad economic and political conditions (such as currency areas and international trade agreements), cultural norms and other factors that provide a contextual framework within which the embedded regime may expand and develop. Landscape conditions can change both suddenly (e.g. the oil 'shocks' of the 1970s) or more slowly (e.g. climate change). What constitutes a niche is a bit vaguer, or more diverse, in the literature on socio-technical transitions. A niche may be a discrete innovation or a cluster of related innovations around which initial markets may become established. But a niche may also be a protected market space (perhaps created by state regulation or the provision of incentives under 'strategic niche management' approaches), and it may be a geographical space (Cooke, 2010) within which new practices and alternative ways of being emerge (as a form of 'grassroots' change).

It is not always clear how a single niche, or a linked cluster of niches, may come to displace an established regime, but there must be some compelling advantages (usually but not exclusively economic) to the adoption of the new practice at least within the niche concerned. Thereafter, and with further development, the emergent niche may simply be better than the incumbent regime along a range of significant parameters, and the regime is largely displaced – though enclaves and remnants may remain. Historically at least, this process has resulted in 'superior' technologies emerging but those technologies (and their associated practices, organizational forms, etc.) have not necessarily been more sustainable. In contrast, the contemporary concern within socio-technical transitions is to achieve progress towards more sustainable regimes but in some important respects the emergent technologies do not embody compelling advantages other than some environmental performance attributes. That is, compared with incumbent technologies and the expectations of performance that these technologies have engendered in social behaviours, market structures and other attributes, the emergent technology may be more expensive, less reliable or offer a reduced suite of performance attributes. Certainly in the realm of cars, pure battery-electric vehicles (EVs) may be quieter and smoother than their internal combustion engine counterparts, with faster acceleration, but have a more limited range, a much longer time needed to take more energy on board, with a limited tolerance for cold or hot climate conditions, more expensive, and in some respects, perhaps, are expected to be less durable (in terms of the battery at least).

If transitions are concerned with change, then inevitably there is an interest in the pace of change. It is surprisingly difficult to establish the appropriate temporal benchmark against which a transition's progress might be measured. Inadequate rates of change then become the focus of analysis: what is preventing change or retarding the rate of change? Where and how might policy be best directed to lubricate friction points, dismantle barriers, in the transitions process? In complex socio-technical systems, is it possible to isolate and act upon individual causal agencies? What happens to pre-existing socio-technical systems? How fast do embedded regimes fade away and what influences this rate of decay? Some research has been directed at these questions, related to the issues of inertia and resistance in socio-technical systems because of the links with concepts such as path dependency (Steinhilber et al., 2013; Turnheim and Geels, 2013; Wells and Nieuwenhuis, 2012), but again the crucial issue of the rate of change desired seems largely absent.

EVs do not just require transition within an established quasi-autonomous socio-technical ensemble (i.e. automotive or automobility regime) but also in related proximate socio-technical ensembles that were previously distinct (i.e. mobile ICT services; electricity generation and distribution) (Roland Berger, 2011) and so the character of interactions between these meta-ensembles may be as significant as those where emergent niche activities destabilize an existing ensemble, in seeking to explain rates of socio-technical change. It might be broadly hypothesized that existing interlocking ensembles are more likely to exhibit inertia, and perhaps in a related manner that contemporary socio-technical systems are likely to be more entangled with each other than has historically been the case, creating a web of mutually reinforcing path dependencies. On the other hand, where a nascent socio-technical ensemble needs a degree of fusion with another ensemble in order to emerge viably, it may be hypothesized that transition is more difficult to achieve because it implies a more challenging task in terms of co-ordination or orchestration.

If the decline phase of the regime cycle is considered, then the critical issue is how long the technologies and social practices as embedded in capital investments, spatial structures, human skills and knowledge, and behaviours and expectations around products in use endure. This is a relevant consideration in particular for the displacement of existing socio-technical ensembles, a process that may be necessary to create the space required for sustainable transitions to occur (the process of creative destruction as it is termed in Schumpeterian analysis). In some cases, pre-existing infrastructures may simply be allowed to wither away, as may be seen in the urban decay that frequently surrounds contemporary life. The canal systems in the UK largely suffered this fate, or were relegated to essentially recreational functions. In other cases, the removal of an entrenched incumbent socio-technical ensemble may be a prerequisite to create the space within which a (more sustainable) alternative is desired. In other instances, it may be that the incumbent organizations that constitute the core of a socio-technical ensemble are themselves (Bergek et al., 2013) the source of transition, or new entrants are not influenced by the destruction of incumbent organizations (Brown et al., 2013). Recent work in this area has started to explore why and how distinct socio-technical ensembles are able to endure or even actively resist change (Turnheim and Geels, 2013; Wells and Nieuwenhuis, 2012). However, this work is somewhat in a vacuum in the sense that there is no clear temporal benchmark against which to measure change or its lack thereof. Hence in seeking an understanding of socio-technical change, it is necessary to combine the analysis of regime creation with that of regime destruction, particularly where the two contesting regimes essentially occupy similar functional spaces. Such a combined analysis is presented in a simple form in this paper. Regime transition is constructed as a process of displacement, and hence we are interested to understand how quickly the entire stock of non-EV cars in use can be removed from circulation under different modelling assumptions. The growth of new EV sales is thus the growth of an emergent (more sustainable) niche, but the impact on the total fleet of vehicles in use is lagged by the rate of turnover of that total fleet, and by expansion in the total fleet before full EV coverage is obtained. In reality, history has shown that older technologies that formed the core of previous regimes tend to persist, albeit in niches. Thus, the much-quoted example from Geels (2002) of the transition from sailing ships to steam was never in fact completed; sailing ships still exist and in fact thrive, providing many firms with a viable market in which to operate, while this technology is in fact still being refined well beyond what was deemed possible in the heyday of sail. Similarly, the decline of the horse economy did not remove horses from society; instead they took on new roles as tools for various leisure pursuits, again occupying a range of viable and sustainable market niches. The analysis also moves on to consider how far EVs could actually contribute to the transition to low-carbon mobility. It is actually this wider concern with low-carbon mobility (or more precisely, sustainable mobility) that transition analysis seeks to speak to (Kemp et al., 2012). Mobility, along with energy generation and distribution, has long been a sphere of analysis for socio-technical transitions.

Transitions theory is largely concerned, then, with the macro shifts in technology and the economy over the long term. Policy formulation, even where it has long-term goals, tends to be narrowly targeted, technologically based and short term in application (Sandén and Azar, 2005). The temporal misalignment between the theoretical framework and the practical implementation of policy therefore represents something of a methodological challenge if the academic content is to have 'impact' in the sense that research funding agencies emphasize nowadays.

2.3 Varieties of Capitalism

The varieties of capitalism school has developed a distinct theoretical discourse within the broader analysis of contemporary business and economics and in so doing has sought to integrate previously rather more distinct threads of discussion in the realms of human resource management, regulation and governance, economic structure and competition into a more nuanced understanding of how capital reproduces itself over time and space. A key focus of earlier work on this theme has been capital–labour relations. Notwithstanding some of the convergence processes that are often subsumed under the rather cavalier title of 'globalization', there remain distinct differences in the capitalisms of place (i.e. in particular that different nation states harbour distinct varieties of capitalism) and distinct differences in the capitalisms of structure (i.e. that state monopoly capitalism is distinct from say the regulated franchise model now found in many countries). Subsequent research has sought a more nuanced understanding for the basis of national differences in capitalism, and indeed the similarities and differences that emerge at different spatial scales from the continental (many nations) to the regional (within a region), drawing upon issues such as political representation and voting structures (Schneider and Karcher, 2010) to enrich the theorization. Out of this work emerges the theme that nation states and other structures that are variable on a nation-by-nation basis such as organized labour have different capacities for engagement and intervention with national and multinational capital. Thus, even if there is a process of global economic integration in some senses, the outcomes are not uniform or indeed simply predictable.

The varieties of capitalism approach therefore seeks insight into similarities and differences on the basis that comparisons of performance or strategy between firms in different places cannot be reduced to a singular or universal rational explanation (Coates, 2005). That is to say, to some degree firms are embedded in their localities, in the cultural and social practices that surround them, and in the institutional, legal and regulatory frameworks that may be more or less specific to place. In an era where some large multinational companies occupy multiple locations and functional activities may be dispersed across those locations, it is a delicate task to tease out an understanding of how far the company may or may not transcend the specifics of place. Early research on globalization tended to emphasize the ability of large companies to benefit from economic liberalization, privatization and the deregulation of formerly protected national markets (Ohmae, 1990). In the context of the automotive industry, there was much interest during this period in the notion of 'Japanization' following waves of foreign direct investment by Japanese vehicle manufacturers into North America and then Europe. This triggered two areas of research interest: first how far these Japanese companies were able to bring with them their practices – particularly in human resource management and supply chain management (Oliver and Wilkinson, 1988); and second how far existing vehicle manufacturers in Europe and North America were able to adopt Japanese (usually conflated to mean Toyota Production System-based) management systems (Womack et al., 1990). From these interests flowed a large volume of research and publishing on issues of convergence, corporate performance benchmarking and the emulation of best practice, approaches that spread far beyond the automotive sector. However, in reaction to this emerged in Europe the GERPISA school of thought, centred initially on French academics, which sought to re-establish the importance of difference. These accounts are more likely to emphasize contestation, the search for alternative solutions and alternative legitimizations, as a counterweight to the 'inevitability' of neoliberalism (e.g. Boyer and Freyssenet, 2000; Freyssenet et al., 1998).

Alternatively, the question of change in distinct varieties of capitalism is also significant, as is the case with Brazil. Contradictory tensions may emerge between forces propelling liberalization, open markets, free movement of capital and the related prescriptions of contemporary neo-conservative economists, on the one hand, and those of populist and national concerns to protect employment and generally control the more rampant aspects of capitalist development, on the other. In this, we echo the critique of Schrank (2009) who questioned the inevitability of right-of-centre governments in Latin America. More radical critiques (Jessop, 2014) have in turn questioned the narrowly institutional focus of the varieties of capitalism approach and argued for 'variegated capitalism' as a temporary and spatially bounded reconciliation of the innate contradictions of capitalism accumulation: a perspective that has gained further currency with the 2009 global economic crisis. While there has been some criticism of the national focus of the varieties approach to capitalism (Lane and Wood, 2009; Walker et al., 2014), particularly with regard to the larger countries such as China (Zhang and Peck, 2014), the determination of the unit of analysis in this sense remains reasonable in the case of Brazil and the issues discussed in this paper because notwithstanding the highly regional dimensions of the sugarcane ethanol business, many of the key institutional and jurisdictional issues are resolved at the national level in Brazil.

One emergent question for the varieties of capitalism school is: how do distinct varieties engage and interact with issues surrounding sustainability (Mikler, 2009)? Almost inevitably such issues result in concerns over how and with what consequence are new regulatory or governance interventions justified and maintained, and how do companies react under those circumstances. The instance of carbon emissions regulation in Europe, for instance, is thus but one discrete example of the wider category of problem: understanding how capitalisms may or may not adapt to new environmental or sustainability imperatives and how the spatial boundaries of governance may operate. Interestingly, there may be a case to argue that sustainability solutions may be grounded in local distinctiveness in a manner that reflects both the specific manifestation of sustainability issues and the spatially constrained nature of the potential solutions. In this regard, then, efforts to improve sustainability at the macro-societal level or the micro-corporate level may serve to reinforce or indeed amplify the distinctions between capitalisms in different spatial jurisdictions.

2.4 Global Value Chains

At first glance, global value chain theory seems antithetical to the ideas contained within the varieties of capitalism concept. Global value chain research seeks to explore and explain what links places together, not what sets them apart. It is a vibrant research area: the website on global value chains (https://globalvaluechains.org) identifies 745 papers on the topic. It takes as a starting point the manner in which previously isolated places, economic activities and social organizations (including companies) have been drawn into a sort of global economic relationship. Much of the research in this area has, in consequence, been concerned with traditional agricultural practices and how they have changed in emerging economies as they are integrated into global chains of production, distribution and sale – usually orchestrated by large multinational companies (Sturgeon, 2013). The other side of the equation, often implicitly, is the literature on 'hollowing out' of production from the mature industrial economies. This is a theme that has preoccupied researchers, trade unionists and policymakers for some time, and more recently has been given renewed emphasis given the rather dubious tax evasion strategies of major companies like Apple, Amazon and Starbucks who seem intent on providing as little as possible to host countries, justifying this on the grounds of optimizing shareholder value.

Global value chain theory is also concerned with networks, rather than the dynamics of individual firms. Not surprisingly perhaps, this theory has been of great interest to geographers concerned with economics and industrial change. A successful firm is one that best orchestrates the supply chain networks it has created, enabling a flow of value creation to pass along. Interestingly, however, the intersections with the varieties of capitalism approach start to emerge with a consideration of forms of governance. One of the leading authors in the global value chains literature, Timothy Sturgeon, has, for example, posited that there are five archetypes of governance in such chains – hierarchy, captive, relational, modular and market (Gereffi et al., 2005) – but the links to socio-technical change and to varieties of capitalism emerge with the idea that differences in value chain governance emerge out of differences in broader institutional, regulatory and societal processes. What is slightly surprising is that the global value chain literature shows scant connection to the burgeoning work on supply chain management, and vice versa. Perhaps, this is more an illustration of the artificial walls created by disciplinary and journal boundaries than any particular animosity between the two areas of research.

Global value chains have also become linked to the idea of local and regional clusters, and of industrial districts. The spatial dimension inside global value chains is therefore rather interesting, because it begins to connect with ideas of geographic specificity. The focus for value chain theory is then of course to understand how spatially distinct clusters of economic activity are integrated into wider structures and processes and via what particular mechanisms. Certainly, this school of thought appears rather less critical of the economic forces sweeping through global systems than might be expected of the varieties of capitalism school: the broad understanding seems to be that less developed locations are beneficiaries of being brought into the monetary economy as part of a development process. However, an overt link with the varieties of capitalism approach is evident because in global value chains the nature of particular locations and the networks within them are crucially important (Sturgeon et al., 2008). Hence, issues such as the degree and kind of unionization, of capital–labour relations, of the role of industry associations, national legal frameworks and even cultural attitudes and norms of behaviour are all important in shaping outcomes in terms of inter-firm governance and supply chain management. Notwithstanding this comment, the global perspective of this theoretical approach inevitably means a focus that goes beyond the boundaries of the individual firm, and beyond the national setting of that firm, and into global networks (Lakhani and Kuruvilla, 2013).

While much of the work has been about commodity products (food and drink, and basic materials; see Dijk and Trienekens, 2012), some has connected with the automotive industry (Humphrey, 2003; Sturgeon and Florida, 2004). A key concern has been the impact that the creation of global value chains has in terms of jobs, trade and product flows around the world economy, but particularly in 'source' and 'destination' countries. In simple terms, as trade has integrated, the global value chains approach suggests that production has disintegrated both spatially and in terms of vertical ownership structures so that 'offshoring' often equally means 'outsourcing'.

2.5 Change in the Automotive Industry: A Synthesis

These three theoretical perspectives on change in industrial systems all have distinct traditions to which they speak, and to some extent are mutually exclusive. Indeed, it could be argued that they are largely contradictory, or at the very least fail to connect in their main concerns. Hence, for example, the varieties approach seeks out distinctiveness and explanations grounded in the possibilities of local uniqueness; while the global value chain emphasizes the imperatives of deeper global economic integration, ceding but little to locality. Moreover, the socio-technical change literature is more concerned with the grand scale of change over decades as technologies permeate society, while the value chains literature is much more about the immediate past and near future, and seeking to understand the volatility of the contemporary economic situation.

On the other hand, there are some interesting synergies that might allow the insights of one of these three theoretical traditions to be enriched by the others. Of the three frameworks, it could be argued that socio-technical transitions is the most 'open' to integrating other theoretical perspectives that might assist in the explanation of the processes and structures that underpin the over-arching system framework. Partly because the socio-technical systems approach is concerned with the bigger questions at a high level of abstraction, and simultaneously with the myriad detail of niche examples, there is a tendency for this innovation literature to intersect with and co-mingle with many other disciplines and theoretical frameworks. Indeed, the basis of the approach emerged out of several distinct traditions including systems theory, evolutionary economics, regulation theory, path dependency theory and more. In this respect, socio-technical transitions insights might be assisted in understanding how and why transitions proceed differentially across space by drawing on the contributions available from the varieties of capitalism literature. This would allow the transitions understandings to be more sensitized to differences at national and regional levels without resorting to the somewhat sterile approach of technology innovation systems.

Similarly, the transitions approach is only just starting to engage with organizational behaviour theories, in part arising out of the realization that incumbent actors can be important not just to resist change, but also to participate in change. In the automotive industry, the global value chains approach has been useful to understand how 'hollowing out' processes occur in mature industrial economies, but the insights from the transitions approach would assist in illuminating the inter-relationships between the permeation of new technologies and the destabilization of market conditions that compels the quest for global value chains as a defensive response. It has been argued by Reji (2013) that global value chains offer an opportunity for small- and medium-sized enterprises to be made more competitive by participating in such chains, which in turn may be one mechanism through which emergent niches can be first protected and then expanded over time.

One problem in the transitions literature is that the strand of thinking on 'strategic niche management' that is concerned with the development of niche activities over time cannot point very strongly to such development processes, with the partial exception of electricity production in Germany. While global value chains speak to the neoliberalism of contemporary economic policy, it is not clear that there is a natural 'fit' with the approaches of both transitions and varieties of capitalism on this important issue. There is much debate within the transitions approach on whether incumbent actors can also participate in (in a more proactive stance) regime transition, or conversely whether the governance and ownership structures of contemporary capitalism would also have to be changed in order for a sustainable transition to be achieved. However, the contribution of the global value chains approach to highlighting issues of power and governance in supplier relations, including in the automotive industry (Sturgeon, 2007), does at least provide some overlap with the concerns of the varieties school of thought. Moreover, for the varieties of capitalism approach, it is notable that global value chains are significant as a mechanism to integrate or at least link diverse jurisdictions and spaces. There are clearly contradictory tensions in the processes that have been pulling more places and people into global economic relations, but it is important to understand further why all locations experience these forces differentially. Hence, global value chains could be a mechanism that acts to increase place diversity while simultaneously bringing greater extension of market economic relations.

2.6 Conclusions

Theory building is an important endeavour in academic social science research precisely because of the character of social phenomena – including the features of an industry such as that which builds and sells cars – which are both enormously complicated and constantly changing. It is indicative that recent work in global value chain theory has called for 'modular' theorization (Ponte and Sturgeon, 2014). This enters a plea for researchers to add new theories, refine theories and add them together as a series of building blocks to enrich our understanding offered by the perspective. A similar call is evidenced in Coates (2005) for the varieties of capitalism approach. The research challenge in this sense is to reconcile two rather contradictory pressures: for simplification and for depth of understanding. Part of the purpose of research is indeed to abstract from reality in order to convey an understanding of the phenomena under investigation, and hence this is a process of simplifying some reality (whether understood objectively or subjectively).

Alternatively, the quest for a theoretically comprehensive and empirically robust explanation may lead to an unending spiral of ever-more elaborate theories and ever-extending empirical reach, alongside burgeoning methodologies and data requirements. Whether there is a grounded scholarship approach, in which theory is elaborated while immersed in practice, or there is a more traditional deductive/inductive approach in which there is a more sequential and linear perspective on theory and empirical research, the innate promiscuity of social science theorization and the manifest empirical richness of a subject as vast as the automotive industry and automobility means that these issues can never be completely resolved.

References

Bergek, A., Berggren, C., Magnusson, T. and Hobday, M. (2013) Technological discontinuities and the challenge for incumbent firms: Destruction, disruption or creative accumulation? *Research Policy*, **42**(6–7), 1210–1224.

Boyer, R. and Freyssenet, M. (2000) *Les modèles productifs*, Paris: La Découverte & Syros.

Brown, J.P., Lambert, D.M. and Florax, R.J.G.M. (2013) The birth, death, and persistence of firms: Creative destruction and the spatial distribution of U.S. manufacturing establishments, 2000–2006, *Economic Geography*, **89**(3), 203–226.

Coates, D. (ed) (2005) *Varieties of Capitalism, Varieties of Approaches*, Basingstoke: Palgrave Macmillan.

Cooke, P.N. (2010) Socio-technical transitions and varieties of capitalism: Green regional innovation and distinctive market niches, *Journal of the Knowledge Economy*, **1**(4), 239–267.

van Dijk, M.P. and Trienekens, J. (eds) (2012) *Global Value Chains: Linking Local Producers from Developing Countries to International Markets*, Amsterdam: Amsterdam University Press.

Freyssenet, M., Mair, A., Shimizu, K. and Volpato, G. (1998) *One Best Way? Trajectories and Industrial Models of the World's Automobile Producers*, Oxford: University Press.

Geels, F.W. (2002) Technological transitions as evolutionary reconfiguration processes: A multi-level perspective and a case-study, *Research Policy*, **31**, 1257–1274.

Gereffi, G., Humphrey, J. and Sturgeon, T. (2005) The governance of global value chains, *Review of International Political Economy*, **12**(1), 78–104.

Humphrey, J. (2003) Globalization and supply chain networks: The auto industry in Brazil and India, *Global Networks*, **3**(2), 121–141.

Jessop, B. (2014) Capitalist diversity and variety: Variegation, the world market, compossibility and ecological dominance, *Capital and Class*, **38**(1), 45–58.

Kemp, R., Geels, F.W., Dudley, G. and Lyons, G. (eds) (2012) *Automobility in Transition? A Socio-Technical Analysis of Sustainable Transport*, New York: Routledge.

Lakhani, T. and Kuruvilla, S. (2013) From the firm to the network: Global value chains and employment relations theory, *British Journal of Industrial Relations*, **51**(3), 440–472.

Lane, C. and Wood, G. (2009) Capitalist diversity and diversity within capitalism, *Economy and Society*, **38**(4), 531–551.

Mazzucato, M. (2013) *The Entrepreneurial State: Debunking Public vs. Private Sector Myths*, London: Anthem.

Mikler, J. (2009) *Greening the Car Industry: Varieties of Capitalism and Climate Change*, Cheltenham: Edward Elgar.

Ohmae, K. (1990) *The Borderless World: Power and Strategy in the Interlinked Economy*, London: Collins.

Oliver, N. and Wilkinson, B. (1988) *The Japanisation of British Industry*, London: Blackwell.

Ponte, S and Sturgeon, T. (2014) Explaining governance in global value chains: A modular theory-building effort, *Review of International Political Economy*, **21**(1), 195–223.

Reji, E.M. (2013) Value chains and small enterprise development: Theory and praxis, *American Journal of Industrial and Business Management*, **3**, 28–35.

Roland Berger (2011) Automotive landscape 2025: opportunities and challenges ahead, Copy obtained from http://www.rolandberger.com/expertise/industries/automotive/2011-02-28-rbsc-pub-Automotive_landscape_2025.html. Accessed 24 March 2011.

Sandén, B.A. and Azar, C. (2005) Near-term technology policies for long-term climate targets – Economy wide versus technology specific approaches, *Energy Policy*, **33**(12), 1557–1576.

Schneider, B.R. and Karcher, S. (2010) Complementarities and continuities in the political economy of labour markets in Latin America, *Socio-Economic Review*, **8**(4–3), 623–651.

Schrank, A. (2009) Understanding Latin American political economy: Varieties of capitalism or fiscal sociology? *Economy and Society*, **38**(1), 53–61.

Steinhilber, S., Wells, P. and Thankappan, S. (2013) Socio-technical inertia: Understanding the barriers to electric vehicles, *Energy Policy*, **60**, 531–539.

Sturgeon, T.J. (2007) How globalization drives institutional diversity: The Japanese electronics industry's response to value chain modularity, *Journal of East Asian Studies*, **7**(1), 1–34.
Sturgeon, T.J. (2013) Global Value Chains and Economic Globalization – Towards a New Measurement Framework, Report to Eurostat. Copy obtained from http://ec.europa.eu/eurostat/documents/54610/4463793/Sturgeon-report-Eurostat. Accessed 10 June 2013.
Sturgeon, T.J. and Florida, R. (2004) Globalization, deverticalization, and employment in the motor vehicle industry, in M. Kenny with R. Florida (eds.): *Locating Global Advantage: Industry Dynamics in a Globalizing Economy*. Palo Alto, CA: Stanford University Press.
Sturgeon, T.J., Biesebroeck, J.V. and Gereffi, G. (2008) Value Chains, Networks and Clusters: Reframing the Global Automotive Industry, MIT IPC Working Paper 08-002. Boston: MIT Industrial Performance Center.
Turnheim, B. and Geels, F. (2013) The de-stabilization of existing regimes: Confronting a multi-dimensional framework with a study of the British coal industry (1913–1967), *Research Policy*, **42**, 1749–1767.
Walker, J.T.; Brewster, C. and Woody, G. (2014) Diversity between and within varieties of capitalism: Transnational survey evidence, *Industrial and Corporate Change*, **23**(2), 493–533.
Wells, P. and Nieuwenhuis, P. (2012) Transition failure: Understanding continuity in the automotive industry, *Technological Forecasting and Social Change*, **79**, 1681–1692.
Womack, J., Jones, D. and Roos, D. (1990) *The Machine that Changed the World*, New York: Free Press.
Zhang, J. and Peck, J. (2014) Variegated capitalism, Chinese style: Regional models, multi-scalar constructions, *Regional Studies,* DOI: 10.1080/00343404.2013.856514.

3

The Market for New Cars

Peter Wells
Centre for Automotive Industry Research, Cardiff Business School, Cardiff University, Cardiff, Wales, UK

3.1 Introduction

From the onset of the global financial crisis in around 2008, it became immediately apparent that any decline in economic growth, or indeed actual recession, has a profound impact on the demand for new cars. It could be argued that this economic event, whose reverberations continued to be felt across the European Union in particular right through to 2014, accelerated and highlighted underlying market trends in any case. Of significance in this regard are the decline of automobility, broadly defined, in the mature industrial economies that first experienced the mass ownership and use of cars; the growth in demand in the markets of India, Brazil, China and much of Asia; and the consequential shift in the 'centre of gravity' of the industry. There is some hint too of a bifurcation in the mature markets as a consequence of the divergence in income and wealth in places such as the United Kingdom and the United States. China has become the largest single national market in the world, and consequently also the most important single market for many companies. New markets may have different priorities and interests, and take the industry in different directions. An intriguing example, if somewhat flawed in execution, is that of the Tata Nano with which pioneered the idea of minimalism in automotive design and production in the modern era. The established markets have to some extent retained their pre-eminence in the luxury and sports segments, and in the emergent electric vehicle applications including car-sharing schemes. Yet at the same time the burgeoning of output, sales and profits elsewhere has helped to underwrite significant shifts in ownership and control that speak to a long-term structural change in the character of the global automotive industry. The takeovers of Volvo by Chinese firm Geely in 2009, and Jaguar Land Rover by the Indian Tata Group in 2008 in particular stand out as important events.

The Global Automotive Industry, First Edition. Edited by Paul Nieuwenhuis and Peter Wells.

3.2 Market Fragmentation and Lack of Industry Consolidation

The automotive industry has long been considered as characterized by the search for production economies of scale as the basis for least cost, both to expand the market and to secure profitability. This accepted truism has, however, proven rather elusive and perhaps illusory. It is by no means certain, for example, that once a company has achieved the status of the highest-volume producer, it will subsequently dominate the market in perpetuity thereafter: the long-run problems experienced by GM despite production pre-eminence, and the more recent difficulties experienced by Toyota with high-profile recalls of defective cars, are illustrative that scale alone at least is insufficient.

One feature that may be of significance is that of market fragmentation. Economies of scale in manufacturing are at their greatest with the highest lifetime volume of a standardized product. While such standardization may be acceptable in rapidly expanding markets new to motorization of the type that underwrote the epic lifetime production volumes of the VW Beetle, it would appear that in broadly saturated markets in which the stock of vehicles in circulation is high, consumers need to be tempted by greater differentiation and variety. The consequence is a reduction in the time for which a product is retained in production, and a proliferation of models, body styles, variants and options that add complexity and cost and therefore further act to undermine traditional economies of scale. Table 3.1 illustrates this issue with respect to the UK market.

In a global market of 70–80 million new cars per annum, there is hypothetically 'room' for five or six manufacturers producing 12–16 million cars per annum and three platforms at 3–6 million cars per platform. Moreover, with the passage of time, the most efficient producers should exclude the least efficient and absorb the market space so vacated. Hence, it is to be expected that the share of the total market taken by the leading five companies is tending towards 100%. Table 3.2 illustrates the global production position with respect to all vehicles in the period 1998–2011.

Some interesting points emerge from Table 3.2. First, the share of total production accounted for by the top 5 and top 10 manufacturers actually declined in the period 1998–2010. Second,

Table 3.1 Brands, body styles and variants in the United Kingdom, selected years

	Brands	Body styles	Variants
1994	54	300	1303
2009	57	376	3637
2013	62	422	3510

Source: Derived from Autocar, various years.

Table 3.2 All vehicles, production by top 5 and top 10 manufacturers in selected years

	1998 (m)	1998 (%)	2008 (m)	2008 (%)	2010 (m)	2010 (%)	2011 (m)	2011 (%)
Top 5	28.6	54.1	33.2	47.9	34.8	44.7	36.7	46.5
Top 10	40.8	77.0	47.9	68.9	51.4	66.1	53.3	67.6
Total	52.9	100.0	69.5	100.0	77.7	100.0	78.8	100.0

Source: Derived from OICA data, various years. http://oica.net/category/production-statistics/

Table 3.3 Market share in Europe 2011 by model (EU-27)

VW Golf	3.6
Ford Fiesta	3.0
VW Polo	2.6
Renault Clio	2.5
Peugeot 207	2.2
Ford Focus	2.0
Fiat Punto	1.9
Renault Megane	1.9
Opel Corsa	1.8
Fiat Panda	1.8

Source: ICCT (2012).

a slight upturn in the share accounted in 2011 was insufficient to compensate for the loss of market share experienced overall. Third, and following on from the 2011 increase in market share for the top 5 and top 10, it is perhaps significant that the global market in this year virtually stagnated compared with the growth from 1998 to 2010. This is suggestive of the ways in which market expansion can defer the process of consolidation and allow differential opportunities to companies depending upon where they are located.

Table 3.3 illustrates that, on a per-model basis, the overall market share even for the market leaders is very modest when looking at the 27 markets of the European Union.

At a global level, there are also enduring differences that mean that there is not just one large undifferentiated market for new cars. As a consequence of myriad regulatory, geophysical, climate, cultural and socio-economic differences, there remains a distinctly national or regional character to many of the global automotive markets and the products developed to serve them despite the apparent ubiquity of some brands and models. Hence, for example, the 'light truck' segment in North America does not really exist as a regulatory category elsewhere; the longstanding preference of North American consumers for large/heavy cars (and light trucks!) with large engines and automatic transmissions has been somewhat diluted as cars elsewhere have grown in size and weight, but still persists as evidenced by the comparatively poor fuel economy performance of the fleet there. In Japan, the opposite applies, with the unique 'Kei' or 'midget' class of microcars (albeit again somewhat tempered for non-Japanese markets) still an important segment, whereas vehicles with engines of more than 2l are something of a rarity. Elsewhere we can highlight the importance of the diesel engine for cars in Europe, of ethanol flex-fuel cars in Brazil and of two- or three-wheelers in much of Asia. Below this, fitment rates of items such as automatic transmissions, air-conditioning, power steering and many other items vary hugely.

Underlying the aggregate changes are some important structural shifts in the automotive industry. In 1998, the global top 5 were GM, Ford, Toyota, VW and DaimlerChrysler. By 2010, the top 5 were Toyota, GM, Volkswagen, Hyundai-Kia and Ford (note Nissan and Renault were listed separately). In contrast, Daimler was down to 12th in 2010 while Hyundai was only 15th in 1998.

At a fundamental level, the explanation for the lack of consolidation is very simple: it is not yet a 'global' industry with a single 'global' market, while diseconomies of scale – particularly at the organizational, human level – may also begin to play a role. It is pertinent to add that it will be a long time, if ever, until such a status is achieved. As a consequence, the larger vehicle manufacturers with traditional market presences in the EU, Japan and North America suffered disproportionately from the recessionary conditions in these regions compared with those based in India and China who were able to benefit from the continuing growth of those markets.

Moreover, at the structural level, the traditional solutions of merger and acquisition have proven problematic in many cases. The established combines of Ford and GM have been largely dismantled, while the DaimlerChrysler merger was over before it really began (albeit at huge economic cost). There has been the occasional success story in this regard, notably of course the relationship between Nissan and Renault and also Hyundai-Kia, but in general it would appear that the process of consolidation is itself extremely difficult, which in turn means that the theoretical benefits of consolidation in terms of cost reduction are difficult to achieve. Hence, while the industry continues to pursue bilateral restructuring (e.g. Fiat with Chrysler), it is much easier for the industry to allow problem companies to disappear. However, given the economic and political status of these companies, it is often extremely difficult for the political system to allow such a demise to occur in a manner that provides effective economic relief to the industry. Whether it be the closure of individual plants, or of entire companies, such rationalization has been strongly resisted both in the European Union and North America.

3.3 Geography of Markets

The geographic shifts in the locus of market growth in the automotive industry are a reflection of and contributor to wider economic changes in supply, demand and economic conditions. Table 3.4 illustrates the broad changes in key markets around the world. What is evident from Table 3.4 is that the established markets within the EU, North America and Japan have over the period 2005–2012 shown substantial fluctuations but no net growth of any significance, compared with the evident growth from China, Brazil, India and others. The growth of the market in China is especially startling: rising from 3.9 million units in 2005 to 15.5 million in 2012. Expansion at this rate is hardly sustainable, as road capacity and other infrastructure cannot be expanded in line with the burgeoning fleet of cars in use.

What is less certain is the time and level at which car sales globally will reach a peak. Urbanization centred on congested mega-cities appears to be the primary prognosis for the future of humanity, alongside greater or lesser scarcity of petroleum fuels. Neither prospect is particularly promising for the continued growth of the automotive industry as currently constituted and around conventional technologies and vehicle concepts.

As Figure 3.1 illustrates, the problem for the automotive industry is not just that new car sales are stagnating and possibly entering long-term steady decline in the traditional markets. It is also that 'motoring' per se is on the decline, leading some commentators to speculate about 'peak car' (Sivak, 2013).

The apparent reduction in automobility requires care when drawing conclusions, as it is too early to say that long-term trends have definitely emerged (Cohen, 2012). It is perhaps significant, however, that the reduction in automobility appears to have started before the 2008 financial crisis that beset many economies, and hence may be indicative of more profound changes in both the demand for mobility and the ways in which that demand is satisfied.

Table 3.4 Geographic distribution of new car sales, 2005–2012

Regions/countries	2005	2006	2007	2008	2009	2010	2011	2012
Europe	*17912308*	*18692262*	*19622391*	*18820893*	*16606947*	*16491307*	*17159553*	*16187240*
EU 27 countries + EFTA	*15551442*	*15882618*	*16064870*	*14823194*	*14487303*	*13792051*	*13601051*	*12537514*
EU 15 countries + EFTA	*14565695*	*14820182*	*14842186*	*13602038*	*13668808*	*12984549*	*12815435*	*11773266*
Austria	307915	308594	298182	293697	319403	328563	356145	336010
Belgium	480088	526141	524795	535947	476194	547340	572211	486737
Denmark	148819	156936	162586	150199	112454	153858	170036	170763
Finland	148161	145700	125508	139669	90574	111968	126123	111251
France	2118042	2045745	2109672	2091369	2302398	2251669	2204229	1898760
Germany	3319259	3467961	3148163	3090040	3807175	2916259	3173634	3082504
Greece	269728	267669	279745	267295	219730	141501	97680	58482
Iceland	18060	17129	15942	9033	2113	3106	5038	7902
Ireland	171742	178484	186325	151607	57453	88446	89911	79498
Italy	2244108	2335462	2494115	2161359	2159465	1961580	1749740	1402905
Luxembourg	48517	50837	51332	52359	47265	49726	49881	50398
Netherlands	465196	483999	504300	499980	387699	482531	555812	502544
Norway	109907	109164	129195	110617	98675	127754	138345	137967
Portugal	206488	194702	201816	213389	161013	223464	153404	95309
Spain	1528877	1634608	1614835	1151176	952772	982015	808051	699589
Sweden	274301	282766	306794	253982	213408	289684	304984	279899
Switzerland	266770	269421	284674	288525	266018	294239	318958	328139
United Kingdom	2439717	2344864	2404007	2131795	1994999	2030846	1941253	2044609
Russia	1520225	1911240	2514920	2897459	1465742	1912794	2653688	2755384
NAFTA	*9221429*	*9301364*	*9045313*	*8230872*	*6569033*	*6833529*	*7363460*	*8639763*
Canada	847436	858826	841585	872720	729023	694349	681956	748530
Mexico	714010	680946	641394	589045	439120	503748	592101	649333
United States of America	7659983	7761592	7562334	6769107	5400890	5635432	6089403	7241900
Brazil	1369182	1556220	1975518	2193277	2474764	2644706	2647250	2851540
China	3971101	5175961	6297538	6755609	10331315	13757794	14472416	15495240
Japan	4748482	4612318	4325508	4184266	3905310	4203181	3509036	4572333
India	1106863	1311373	1511812	1545414	1816878	2387197	2510313	2773516
South Korea	941483	977140	1040372	1020457	1234618	1308326	1316320	1293585
South Africa	419868	481558	434653	329262	258129	337130	396292	440002
All countries	*44755211*	*47495279*	*50325808*	*49481832*	*48996321*	*55118246*	*57242673*	*60486524*

Source: Adapted from OICA, various years. http://oica.net/category/sales-statistics/

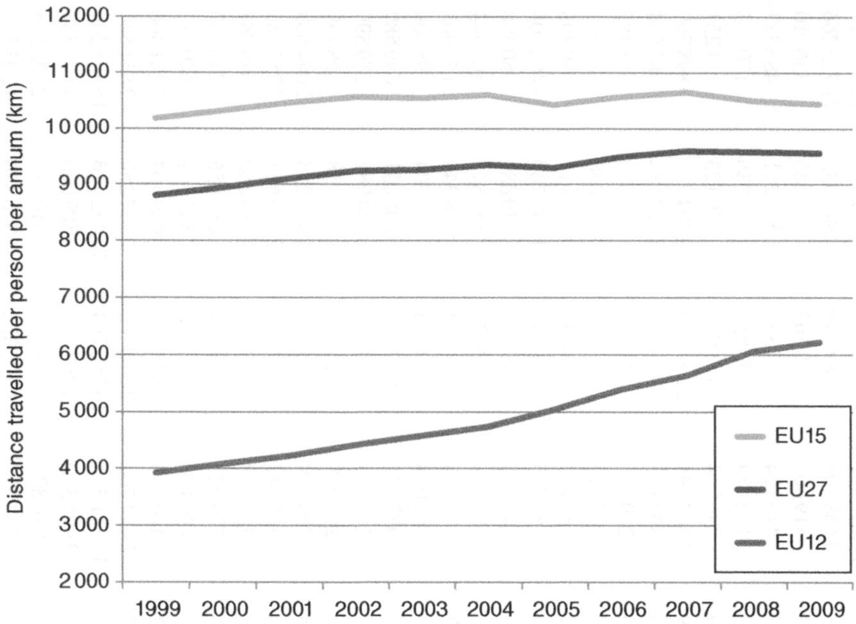

Figure 3.1 Average per capita distance travelled, European Union 1999–2009 (Source: Eurostat, various years)

One example cited of emergent post-automobility is the reduction in the proportion of young people (say under 25 years old) with a driving licence, or owning or having access to a car (Kuhnimhof et al., 2012). Given that populations are on average also ageing in many post-industrial countries, the decline in the proportion of young people driving is clearly important for long-term market demand for new cars. How far the fall in qualified drivers in the young is attributable to disinterest, apathy, rising cost or outright opposition to driving has not really been established. Certainly, the idea here is that if the young do not embrace automobility now, they are less likely to embrace it later in life also. On the other hand, there is some suggestion that there are deeper structural forces at play here. One notable instance is that of urbanization or, in effect, a return to the city and a possible decline in the attractiveness of suburban, as opposed to urban lifestyles. In the United Kingdom, according to the RAC (2013), in 2012, there were 35.2 million licensed drivers (72% of people aged 17 or over), an increase from 31.4 million in 2000. Some 25% of all households do not own a car, but 32% have two or more. Of those households in the lowest 20% of the population by income, 49% do not have a car. Hence, the penetration of automobility has always been partial even in relatively heavily motorized countries such as the United Kingdom, and typical excluded populations include the poor, the young, some of the old and the disabled.

It is also worth noting that, in contrast, the older generations appear to retain cars for longer than was hitherto the case, and the continued use of a car is taken as a key indicator of personal welfare (Gagliardi et al., 2010). Vehicle manufacturers are designing cars for the elderly (Herriots, 2005), and many innovative technologies are important enablers of continued motorization for the elderly. It is then the elderly generations that have the disposable income and predisposition to car ownership and use, but for the automotive industry the concern must be that new recruits into the market will be harder to find.

The reduction in miles driven per annum is possibly an indication that the utility of driving has decreased. Proximate explanations for this decline in utility could include, for example, increased congestion, uncertainty over trip times, poor road conditions, reduced parking opportunities at greater cost (a parking bay in Boston with space for two cars was sold for US$560,000 in 2013; see BBC, 2013), car-free zones, and increased fuel, taxation and insurance costs. Innovations such as social media sites can substitute for some driving trips (Sivak and Schoettle, 2012), while low-cost airlines compete with cars over longer distances, as do home delivery services from supermarkets for shorter distances. New car sales may not directly reflect the reduction in automobility, however, because the real purchase cost of cars, adjusted for inflation and for the enhanced features of successive generations of car, has fallen considerably over time (Wells et al., 2013), so it becomes less costly to keep a relatively under-used vehicle on the driveway.

In some circumstances, car-sharing schemes (see Späth, 2012) may also offer a scenario in which individuals forego the personal ownership of cars, but rather access them through a shared pool of vehicles or other mechanisms such as peer-to-peer rental and car commute sharing schemes (Cairns and Harmer, 2011). In theory, fewer cars are required for a given population compared with the situation where cars are individually owned (Firnkorn and Müller, 2011, 2012). It is worth noting that under individual ownership a car can remain not in use for around 95% of the time, but with car-sharing schemes the intensity of usage is much greater. In turn, this may result in a higher replacement demand. Both vehicle manufacturers and the traditional daily rental sector have become interested in participation in car-sharing schemes (see, e.g. the Daimler project Car2Go and the acquisition by Avis in 2013 of the Zipcar business). What is less certain is how far this type of mobility market represents a substantial commercial opportunity for the future.

The focus on the functionality of the car may to some degree be misplaced. While not a subject that has been given a great deal of research attention, it is intriguing to note how frequently motoring is neither efficient nor economic as an individual mode choice, but remains the preferred option. As a consequence, people express the 'need' for a car in the face of obvious financial or functional challenges. Loderick (2012) reports on research by a UK insurance company to the effect that:

> Almost a third of motorists are sacrificing food in exchange for keeping their cars on the road, according to new research.... The savings motorists are making on food are being used to fund rising motoring costs such as fuel and car insurance.

It is worth noting that some 44% of respondents in the survey also said that they had reacted to rising motoring costs by seeking to use their cars less, but either because of necessity or preference the prioritization of the car is remarkable. Part of the explanation may reside in the path dependency created around the spatial separation of work, housing, retail opportunities, leisure, schooling and so on in a manner not conducive to public transport. Part of the explanation may equally reside in the 'cocooning' character of the modern car as a modest personal space in a crowded world (Bijsterveld, 2010). Certainly, the emotional and psychological appeal of the car transcends mere functionality, and as such suggests the personal car ownership and use may be rather more enduring than some have forecast (Nieuwenhuis 2014; Wells and Nieuwenhuis, 2013).

3.4 Mobility Services and the Emergent Automotive Ecosystem

It is fashionable nowadays to talk in terms of the 'automotive ecosystem', especially for consultants and analysts eager to sell their knowledge and insights into this brave new world of mobility services (Accenture, 2011; Dammenhain and Ulmer, 2012; E&Y, 2011; Rishi et al., 2008). Consultants such as Oliver Wyman (2010) claim that 'mobility solutions' are imperative if customers are to be reached profitably by the established vehicle manufacturers. The underlying idea behind the automotive ecosystem idea is very simple. Rather than selling customers a car and leaving them to get on with the business of using it, vehicle manufacturers will become part of a constellation or ecosystem of inter-linked products and services that combine to deliver mobility packages. More academically, this is a version of network value creation around new business models as discussed in Chapter 18.

Much of the expectation with regard to mobility services hinges upon the uptake of alternatives to traditional petrol and diesel to power the car. However, the rate and extent of penetration of such alternatives is in severe doubt, as is discussed further in Chapters 16 and 17. Despite the structural problems facing the automotive industry, and global economic conditions, there remains a broad consensus that (i) global vehicle sales and total vehicles in use will continue to grow to at least 2030 and (ii) the majority of those vehicles will continue to use petrol or diesel power rather than any of the alternatives. For example, BP (2012) forecast that the global fleet of vehicles will grow by 60% to 1.6 billion by 2030, mostly outside the OECD, and that petrol, diesel or biofuel will power 94% of those vehicles, compared with 4% for natural gas and only 1% for electricity (note that both current hybrids and plug-in hybrids still feature an internal combustion engine). The BP report highlights the lack of policy support and infrastructure as constraining the growth in share of alternatives.

The ecosystem is comprised differently according to who is undertaking the analysis and can be drawn more or less widely. In a narrow sense, there is the convergence between the existing industries in the automotive, computing, mobile telecommunications and electricity distribution sectors around the emergent mobility services for battery-electric cars. Beyond this core, however, are many other potential participants including the daily rental industry, the public transport sector, software suppliers, mapping companies, infrastructure providers and managers, traffic management authorities, financial systems suppliers, electricity generators, vehicle service and maintenance sectors, and others. It is held that companies need to form alliances or networks of association in order to bundle an appropriate package of product and service in an attractive manner to potential consumers, but in practice there is scant agreement on what sort of alliances or networks, or what sort of product-service bundle, will really work. Rishi et al. (2008) define these ecosystems as being automotive, energy, consumer electronics, communities, geographies, social networks, other industries (software, telecommunications and financial services) and government. Dammenhain and Ulmer (2012) take a view that is narrower, but more detailed. In their framework, the e-mobility ecosystem comprises vehicle manufacturers, suppliers to them, the electric vehicles, the IT provider, the e-mobility technology supplier, the e-mobility provider, the public sector, the utility, the distributor and the battery charging/changing operator.

Governments are seeking to support 'pre-competitive' clusters to promote alternative technologies, but given constraints on budgets and the relative failure so far of both fuel cell vehicles and mainstream battery-electric vehicles to really establish themselves in the market, governments and their constituent agencies and departments are perhaps nervous about underwriting these new markets any further. High-profile government-funded failures such as A123 Battery Systems (in 2013) have further undermined the cause of the alternative automotive ecosystem concept as a future market.

Questions are beginning to emerge about the basic character of our approach to automobility. In particular, it would appear that the resource-intensive, over-specified and almost baroque contemporary car is seen as increasingly inappropriate for many urban applications. With dual concerns over global climate change and fuel scarcity, the regulatory pressures on contemporary car design are certainly increasing. The EU, the United States, China and Japan have all enacted regulations on vehicle fuel economy, albeit of varying stringency and with varying penalization for non-compliance. The research focus of the future therefore may be towards non-car alternatives including very light urban vehicles based on the quadricycle concept, electric two-wheel transport, enhanced public transport and alternatives to mobility per se.

3.5 Conclusions

Now, more than ever, the automotive industry faces uncertainty over the future of markets. Contradictory trends appear to be at work. The industry is pushing for ever greater global integration, for example via the use of the UN standards system to harmonize vehicle design and through increased inter-corporate integration. Yet at the same time, markets appear to continue to fragment, and the core concept of the petrol/diesel car is under threat from alternatives more in tune with the economic and environmental concerns of the era. While travel behaviours appear rather obdurate and resistant to change, and the consumer status of the car has not entirely disappeared, there are certainly signs that in the established markets the use of the car has peaked. In the emergent markets that peak has not yet arrived, but it will arrive sooner (and as a peak be lower) than that experienced in the densely motorized countries.

References

Accenture (2011) Changing the game: Plug-in electric vehicle pilots, http://www.accenture.com/SiteCollectionDocuments/PDF/Accenture_Utilities_Study_Changing_the_game.pdf. Accessed 6 September 2011.

BBC (2013) Boston woman pays $560,000 for two parking spots, http://www.bbc.co.uk/news/world-us-canada-22910579. Accessed 14 June 2013.

Bijsterveld, K. (2010) Acoustic cocooning how the car became a place to unwind, *Senses and Society*, **5**(2), 189–211.

BP (2012) BP energy outlook 2030, http://www.bp.com/liveassets/bp_internet/globalbp/STAGING/global_assets/downloads/O/2012_2030_energy_outlook_booklet.pdf. Accessed 1 October 2012.

Cairns, S. and Harmer, C. (2011) *Accessing Cars: Insights from International Experience*, London: RAC Foundation.

Cohen, M.J. (2012) The future of automobile society: A socio-technical transitions perspective, *Technology Analysis and Strategic Management*, **24**(4), 377–390.

Dammenhain, K.-O. and Ulmer, T. (2012) Managing the change to e-mobility, Capgemini, http:\\www.capgemini.com. Accessed 25 March 2015.

E&Y (2011) Beyond the plug: Finding value in the emerging electric vehicle charging ecosystem, Ernst and Young, http://www.ey.com/Publication/vwLUAssets/Finding_value_in_the_electric_vehicle_charging_ecosystem_pdf. Accessed 17 April 2015.

Firnkorn, J. and Müller, M. (2011) What will be the environmental effects of new free-floating car-sharing systems? The case of car2go in Ulm, *Ecological Economics*, **70**(8), 1519–1528.

Firnkorn, J. and Müller, M. (2012) Selling mobility instead of cars: New business strategies of automakers and the impact on private vehicle holding, *Business Strategy and the Environment*, **21**(4), 264–280.

Gagliardi, C., Marcellini, F., Papa, R., Giuli, C. and Mollenkopf, H. (2010) Associations of personal and mobility resources with subjective well-being among older adults in Italy and Germany, *Archives of Gerontology and Geriatrics*, **50**(1), 42–47.

Herriots, P. (2005) Identification of vehicle design requirements for older drivers, *Applied Ergonomics*, **36**(3), 255–262.

ICCT (2012) Databook, International Council on Clean Transportation, http://www.theicct.org/sites/default/files/publications/Pocketbook_LowRes_withNotes-1.pdf. Accessed 12 June 2013.

Kuhnimhof, T., Buehler, R., Wirtz, M. and Kalinowska, D. (2012) Travel trends among young adults in Germany: Increasing multimodality and declining car use for men, *Journal of Transport Geography*, **24**, 443–450.

Loderick, N. (2012) Thousands cut food spending to keep cars on road, http://www.confused.com/car-insurance/articles/thousands-cut-food-spending-to-keep-cars-on-road. Accessed 28 November 2012.

Nieuwenhuis, P. (2014) *Sustainable Automobility: Understanding the Car as a Natural System*, Cheltenham: Edward Elgar.

Oliver Wyman (2010) The future belongs to electric vehicles, Automotive Manager 1/2010.

RAC (2013) Keeping the nation moving, RAC Foundation, http:\\www.racfoundation.org. Accessed 13 June 2013.

Rishi, S., Stanley, B. and Gyimesi, K. (2008) *Automotive 2020: Beyond the Chaos*, Somers: IBM Institute for Business Value.

Sivak, M. (2013) Has motorization in the US peaked? Part 2: The use of light-duty vehicles, UMTRI-2013-20, University of Michigan: Transportation Research Institute.

Sivak, M. and Schoettle, B. (2012) Recent changes in the age composition of drivers in 15 countries, *Traffic Injury Prevention*, **13**(2), 126–132.

Späth, P. (2012) Appreciating incumbent's agency: The case of a local emobility initiative in Stuttgart, Germany. Paper presented at IST2012 Conference, Copenhagen.

Wells, P. and Nieuwenhuis, P. (2013) From 'Freedom of the open road' to 'Safe in the city': Understanding the enduring appeal of personal private automobility. Paper presented at the SCORAI Conference, Clark University, 14–16 June 2013.

Wells, P., Varma, A., Newman, D., Kay, D., Gibson, G., Beevor, J. and Skinner, I. (2013) Governmental regulation impact on producers and consumers: A longitudinal analysis of the European automotive market, *Transportation Research A: Policy and Practice*, **47**, 28–41.

4

Understanding People and Cars

Lorraine Whitmarsh and Dimitrios Xenias
School of Psychology, Cardiff University, Cardiff, Wales, UK

Individual travel choices and vehicle preferences play a key role in shaping the automotive industry inasmuch as this industry is seen as market-driven, which may not always be the case as discussed in Chapters 3 and 12. Here, we provide an overview of the literature on travel and vehicle choices (Sections 4.1 and 4.2, respectively), and on the acceptability of transport policies and new technologies (Section 4.3). A broad range of factors influence individuals' decisions and behaviour within the transport context and their responses to changes to this context. These factors include modal choice, transport demand and choice, meanings attached to cars, public attitudes to transport policies, policy acceptability and attitudes to new transport technologies, including electric vehicles (EVs). We stress the complex interaction of factors shaping transport-related attitudes and behaviour, and the challenges associated with changing these behaviours. Critically, in focusing on relationships between 'people' and cars, we are interested not only in people as consumers but also as citizens and members of social groups (families, organizations and communities). This reframing indicates a broader role for the public than merely as purchasers of vehicles, to embrace decision-making within various private and public (including policy-making) contexts.

4.1 Influences on Travel Choices

Although income is a significant predictor of travel demand (see Figure 4.1), travel decisions in transport are considerably more complicated than the conventional rational economic choice model of behaviour would suggest (Köhler, 2006; Newman, 2013; Small, 2001). For instance, contemporary urban form has developed around roads and cars (cf. 'carchitechture'; Schiller et al., 2010) and has created a strong lock-in to automobiles as the primary form of personal transport in wealthy societies, to the extent of severely restricting – by law,

The Global Automotive Industry, First Edition. Edited by Paul Nieuwenhuis and Peter Wells.

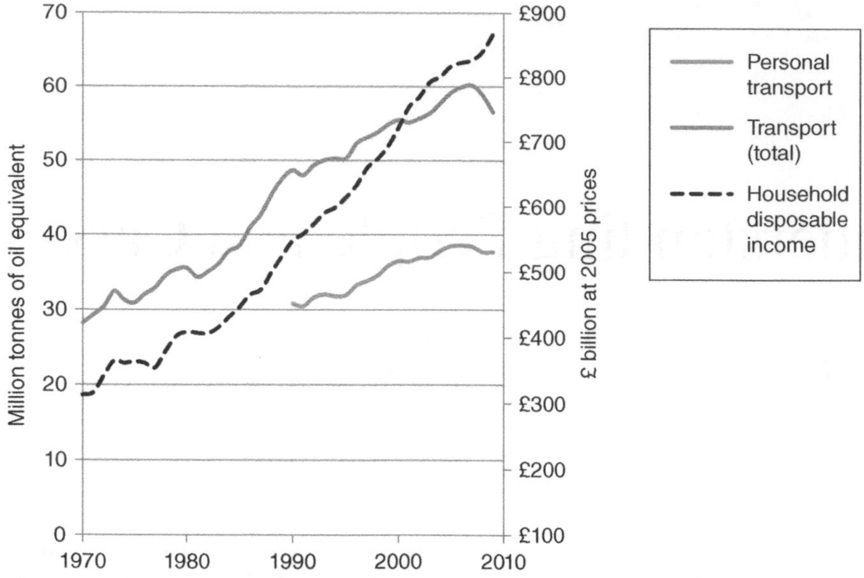

Figure 4.1 Rising UK travel demand and income (Based on DECC, 2010)

in some cases – one of the most natural human activities: r5 that of walking (Moran, 2006). Furthermore, transport behaviour may be a product of multiple motivations or of unconscious factors (e.g. habit, cultural and social norms). Personal preferences for comfort, convenience, autonomy and so on clearly play a role in transport choices (e.g. Whitmarsh et al., 2009), as do less conscious determinants, such as social identity, symbolism and status associated with vehicle choice and use (e.g. Graham-Rowe et al., 2012; Newman, 2013; Schuitema et al., 2013; Steg et al., 2001). Income and pricing of transport options are also important (e.g. Goodwin et al., 2004), as are infrastructure and availability of alternatives (Köhler, 2006); for instance, those living in rural areas are most likely to mainly rely on their cars because of the lack of alternatives in the rural context (DEFRA, 2002; Xenias and Martineau, 2012).

Consequently, *predicting* travel behaviour is a complex matter, and research has employed a number of methods in an attempt to reduce this complexity. Segmentation models, for example, have identified distinct types of travellers (e.g. Anable, 2005; Thornton et al., 2011) based on habits, identity, values, demographic factors, personality, social awareness and so on. In the UK context, Anable (2005) identified six distinct clusters based on analysis of attitudinal data (including attitudes to car use, alternatives, environment and green behaviour) in respect of modal choice, ranging from 'Die-hard drivers' to 'Car-less crusaders'. Four of the six segments were car-owning and accounted for 93% of the sample (in Thornton et al., 2011, car owners account for six of nine segments and 82% of the population), but the segments varied considerably in terms of vehicle attachment, attitudes to environmental impacts and perceived behavioural control over alternatives. A noteworthy segment here is that of young urbanites who choose not to own a car. This represents 7% of the population in Thornton et al. (2011) and also a growing global tendency of younger people driving less

(Polzin et al., 2011). If this trend continues, this will be the first time a generation owns fewer cars and drives less than their parents.

Although most people in developed nations (and increasing proportions in developing countries) own cars, car *use* varies widely in different contexts. The car is still the dominant mode of transport for short distances, and where individuals choose to switch to low-carbon alternatives to driving, this is more often out of a desire to save money or for reasons of convenience or health benefits than out of environmental concern; the latter is, however, often cited as a secondary reason (Whitmarsh, 2009). Indeed, while public awareness of problems with the transport system (e.g. pollution, noise and greenhouse gas emissions) is widespread (e.g. Lethbridge, 2001; Whitmarsh et al., 2009; Xenias and Whitmarsh, 2013), there are various barriers to changing lifestyles. Barriers in turn prevent public awareness manifesting in behaviour change (cf. the well-established 'value action gap'; Barr, 2004; Blake, 1999). One such barrier is often thought to be the lack of knowledge of the relative impacts of different transport modes. Although this may be one factor, knowledge deficit is by no means the most important (Whitmarsh et al., 2009; Xenias and Whitmarsh, 2013). Institutions and infrastructures solidify car dependency alongside other aspects of carbon-intensive lifestyles. Social norms and cultural beliefs serve to reinforce the assumption that car ownership is a precondition of quality of life, which indirectly reinforces the perceived value of automobility (Newman, 2013; Urry, 1999). At the same time, the built environment has developed around – or was altered to accommodate – the car and in so doing perpetuated car dependence (for instance, by covering tramway tracks with asphalt, or cutting busy roads through previously quiet neighbourhoods). Such measures contribute to widespread perceptions of limited (or unattractive) alternatives to driving (Lorenzoni et al., 2007; Lyons et al., 2008). Limiting alternatives or making them less attractive is an established route into path dependence (Arrow, 2000; Kay, 2005). This leads to a self-fulfilling prophecy (e.g. people drive more in the absence of good alternatives, which discourages the development of better alternatives *because* people prefer to drive). This in turns leads to a 'lock-in' (cf. Vergne and Durand, 2010) in automobility.

The term 'behavioural lock-in' has also been coined (Jackson, 2005) to describe the role of habits in restricting desirable lifestyle change: behaviours that are mainly habitual are difficult to change. Travel behaviour, and specifically driving, is often habitual: individuals with strong car-use habits do not consciously deliberate over travel choices or pay attention to information about alternative modes (Verplanken et al., 1997). This clearly works against the effectiveness of information campaigns, which, however, have been the staple of government interventions for a variety of issues. Furthermore, where car use becomes a strong habit, individuals tend to exaggerate the poor quality of alternatives (Fujii et al., 2001). Given this background, it is unsurprising that past informational and economic approaches to encouraging transport behaviour change have met with limited success. Information will be ignored in the presence of strong habits, and economic motivations are only one amongst many reasons for transport choices. Indeed, where economic measures are inappropriately applied, they can lead to public protests, as in the case of the fuel duty protests in the United Kingdom in 2000, where hauliers blockaded oil refineries leading to major disruption and subsequent economic damage (BBC, 2000; Harrabin, 2004; Hetherington and Ward, 2000; Hetherington et al., 2000).

There also needs to be some *motivation* for changing behaviour. This may be intrinsic or extrinsic. Environmental concern may be enough to encourage some individuals to change, although often this will not occur until they are reconsidering travel options for some other reason (e.g. due to moving house, or changing job – also called 'moments of change'; see Verplanken and Wood, 2006). One relevant finding, for instance, is that environmentally conscious people are more likely to travel to work using slow modes or public transport if they have recently relocated, compared to environmentally conscious people who have not moved, and therefore may have strong driving habits (Verplanken et al., 2008). Often, the encouragement to change behaviour comes from external incentives, penalties or restrictions – for example, a free 1-month bus pass, parking restrictions or closure of a highway (Bamberg, 2006; Fujii et al., 2001). However, depending on their duration, such interventions to encourage behaviour change may not be enough to effect long-term change and behaviour change may become extinct once the intervention is concluded. Evidence suggests it takes around 3 months for a new habit to become established (Lally et al., 2010), longer than many behaviour change interventions. At the same time, interventions tend to work best when targeted to 'moments of change': moments in time when individuals are reconsidering their transport choices. This may be the result of significant life changes, such as when relocating or changing job, when people are reorienting themselves in a new situation and are therefore more open to 'unfreezing' their habits: 'habit discontinuity' (Verplanken and Wood, 2006).

In general, however, there is significant resistance to changing travel habits. This finding is consistent across countries (e.g. for the United States: O'Connor et al., 2002). In a UK survey, for example, a notable 50% stated that they 'would only travel by bus if I had no other choice'. Furthermore, only 23% agreed that 'for the sake of the environment car users should pay higher taxes' (DEFRA, 2009). Commonly cited reasons for not changing travel habits include perceived inconvenience, unavailability or unattractiveness of public transport, and safety concerns about alternatives, such as risks of a traffic accident in the case of cycling, or concerns about personal safety in the case of public transport (Black et al., 2001; Davies et al., 1997; Emmert et al., 2010). There is more willingness to consider modifications of trip patterns than changing mode of transport used (King et al., 2009). Finally, in contrast to domestic energy use, where there is more willingness to save energy, few people feel socially obliged to reduce car use (Anable, 2005; Emmert et al., 2010; King et al., 2009). Indeed, there are clear cultural and social associations with travel behaviours and modes; in particular, many enjoy the flexibility, autonomy, comfort and privacy of travelling by car, features which are not offered by car-sharing or public transport (Emmert et al., 2010; King et al., 2009). Some of these barriers were targeted by EU-funded projects focusing on electric vehicles in shared fleets (eBRIDGE, 2013; Niches, 2009).

In sum, then, there are several factors affecting travel mode choices; some being more influential than others. These factors will weigh differently in different contexts, for example in rural areas where public transport is indeed poorer than urban public transport. To be sure, automobility has an important role in transportation and, used appropriately, it can fill important gaps in the transportation fabric, for instance in the case of extra-urban regional – rather than inter-city – travel where the maintenance of frequent and reliable public transport networks may be unfeasible or uneconomical. On the other hand, contemporary cars – principally designed for long-distance travel (Mitchell et al., 2010) – may not be an optimal travel solution when they compete with better suited alternatives, such as urban public transport, which provide more efficient transit within cities. On balance, travel choices are, and can be, influenced by social, cultural, psychological and infrastructural factors and their relative influence can vary with reference to the situation in which a trip is planned.

4.2 Influences on Vehicle Choice

Similar to travel choice, car ownership is predicted by income as well as several other socio-economic factors, including age, household size and the number of drivers in the household. There are additional factors such as the driver's lifestyle and personality (McGeevor, 2009). From a psychological perspective, it is noteworthy that the choice to own and run a car is often *not* an economically rational one – contrary to long established views pervading some policy circles: drivers often underestimate the costs of running a car relative to other modes, considering only the fuel costs of each journey and ignoring the fixed running costs (see McGeevor, 2009) and/or because car owners feel obliged to use their expensive asset, after purchasing. Indeed, driving is often undertaken because it is intrinsically enjoyable and is hedonically valued, and research has shown car use can afford psychological benefits in some cases (e.g. Hiscock et al., 2002; see also Nieuwenhuis, 2014). The choice of a particular vehicle is similarly predicted by both economic and non-economic factors. For example, low income, younger, female and better educated consumers seem to be more likely to drive small cars (Choo and Mokhtarian, 2004). Initial outlay has been shown to strongly influence car-buying decisions for many consumers, which helps explain the success of the scrappage policy in the United Kingdom and Germany (Lane, 2009). However, while consumers often cite instrumental reasons, such as price or safety, for vehicle choice, research that taps unconscious motivations reveals that social identity, symbolism and status are also important influences (e.g. Steg et al., 2001).

Cars are very often symbols of 'conspicuous consumption' (Braun and Wicklund, 1989; Dittmar, 1992). This was first described by Veblen (1899) who observed that for certain goods, the classic 'law of demand' does not apply but is reversed: consumers would buy a good car *because* it is expensive, in order to signal wealth and power. More recently, it has been established that car ownership is also used to communicate aspects of individuals' personality: for example, certain types of vehicle are more associated with masculinity or traits (e.g. aggression) than others (Walker, 2013). Equally, energy-efficient vehicles may be adopted to communicate aspects of drivers' identity or values – in particular, relevant to the environment. One relevant US survey found consumers bought a hybrid vehicle primarily in order to demonstrate their green credentials rather than because of fuel efficiency or low emissions (CNW Marketing, 2007; cited in Maynard, 2007). Indeed, actual environmental impact is seldom prioritized in the car-buying process (McGeevor, 2009). Conversely, social factors – such as the brand of vehicle purchased by neighbours – play an important role in the car-buying process, and indeed in other consumer behaviours. Summarized in plain English as 'keeping up with the Joneses', this psychological mechanism has its roots in social comparison theory (Festinger, 1954) and acts as a 'heuristic' or shortcut (i.e. cognitive rule-of-thumb) to help simplify the complex decision-making process (Bakken, 2008). Other heuristics include brand loyalty, whereby consumers often buy vehicles made by the same firm (or country) due to positive past experiences – a concept that is familiar to the automotive industry (e.g. Mannering and Winston, 1991; Train and Winston, 2007). The complex interplay of such social and experiential factors results in limiting the impacts of any economic incentives (e.g. vehicle duty) on car-buying behaviour (McGeevor, 2009). However, the context is once more an important consideration, and these impacts vary by country and consumer group: emission-based vehicle taxation introduced in Denmark had a significant effect on vehicle emissions (TNO, 2006), while in the United Kingdom it has had minimal impact on consumers' purchase decisions, apart from fleet buyers, who, fortunately, constitute the majority of new buyers in this market (UKERC, 2009).

4.3 Acceptability of Transport Policies and New Technologies

Both the choice of travel and vehicle exert significant influences on the automotive industry; however, they do not exist in a vacuum. As mentioned earlier, context is also important, and we will now turn to the policy and technology contexts. The introduction of new technologies and policies to change travel choices come and go, but they invariably require public support to be successful. First, in respect of new technologies within market-driven economies, public demand for new products is required for market success. A less visible, but equally important point regards public acceptance of any associated infrastructure (e.g. EV charging points and supply networks), which is also necessary. For example, European policy (European Commission, 2011) foresees conventional vehicles being phased out by 2050, and replaced with new (e.g. electric) vehicles and charging infrastructure. Lack of public demand for and support of these new technologies and associated infrastructure would severely hamper progress towards these 2050 targets. Second, in respect of demand reduction and modal shift, significant behavioural shifts are implied; achievement of such behaviour shifts requires public acceptance of economic, land use and associated policies. For example, across Europe, a 50% modal shift from road to rail is proposed for medium distance travel by 2050 (European Commission, 2011), with significant implications on a social and individual level. In such conditions, citizens can have an active role in informing and improving policy, leading to better and more acceptable decisions (Dietz and Stern, 2008; Siebenhüner, 2004). Given that citizens are a key stakeholder group for transport decision-making, we have argued they warrant greater attention in transport policy and technology research (see Xenias and Whitmarsh, 2013).

Although there is broad public consensus that action should be taken to tackle transport problems, most see the government rather than individuals as responsible for taking action (Whitmarsh et al., 2009; Xenias and Whitmarsh, 2013). Consistent with this, there is more support for new technologies or policies to encourage behaviour change (e.g. improved public transport) – termed 'pull measures' (Eriksson et al., 2008) – than 'push measures', such as increased taxes or tolls which might restrict individual freedom (Anable et al., 2006; King et al., 2009). In general, citizens prefer transport policies that are perceived to be fair, effective and which do not limit freedom (e.g. King et al., 2009; Jaensirisak et al., 2005; Schuitema et al., 2010). In many countries, there is lack of support for congestion charging (e.g. Scottish Government, 2009), for example, although attitudes have become more positive in London since the introduction of the congestion charge, suggesting experience of a successful scheme can change opinion (Richards, 2006). A critical factor here is transparency regarding the proposed changes, use of revenues and similar information that build public trust and help acceptance (Schlag and Teubel, 1997). Similarly, King et al. (2009) found that willingness to change travel behaviour became more positive after deliberation and particularly following consideration of personal benefits (e.g. health, financial and time) accruing from combining trips, eco-driving or using more local facilities.

Our research on public attitudes towards different transport options highlights disparity between the public (i.e. transport users) and transport experts (i.e. those working in the field of transport research, policy and implementation) in respect of attitudes to transport policies and technologies. This distinction is significant because it delineates those voices which tend to inform policy (i.e. 'experts') from those who tend to be impacted by policy ('public'). On the one hand, Whitmarsh and colleagues (2009) found that both groups agreed on the need to address problems of unsustainability in the transport sector, and identified broadly similar environmental, social and economic criteria for sustainable transport; on the other hand, amenity of transport was found to be more important for public participants, while expert participants focused more on pragmatic and technological issues. Gerike et al. (2008) also found not only that the public and policy-makers favoured modal shift (e.g. promotion of cycling) but also that both groups considered demand management measures (e.g. parking management) unacceptable.

The reasons for the (partial) disparity between experts and public are multiple and include psychological, social and institutional factors (for a review, see Whitmarsh et al., 2011). Some studies focus on the different types of information processing that experts and non-experts apply. According to this view, in everyday life, our judgements are often affective, automatic and rapid ('experiential reasoning'; Weber, 2010), whereas scientific and technical assessment is more deliberative, conscious and cognitively based ('analytic reasoning'; Weber, 2010). However, this interpretation has been regarded as problematic because it may favour scientific judgements and imply that non-experts are 'irrational' or 'wrong'. Rather, the two groups may use *different* criteria for assessing alternative options, with the public considering more local and contextual factors than experts (Wynne, 1996). Our research (Xenias and Whitmarsh, 2013) has explored the reasons for expert-lay disparity in responses to transport innovation and policy. We compared attitudes of experts and British citizens using focus-group discussion, open-ended questionnaires, attitudinal scales, analytic hierarchy process and preference ranking. We found that both samples prioritized reduction in transport demand in qualitative (open-ended survey) measures. In quantitative measures (where pre-defined options were provided), however, experts preferred techno-economic measures while citizens prioritized behaviour change and public transport improvement. This is an important distinction because expression of preference is partially related to the method of enquiry: pre-defined questions/options relate to top-down information processing, while open-ended options relate to bottom-up processing, and therefore more spontaneous expression. Some options for sustainable transport also correlated with individuals' environmental values, suggesting that expertise alone did not fully account for variation in attitudes. Different perspectives and values imply a need for a broader definition of expertise in transport policy-making, and that the public may not accept transport policies/technologies designed by expert-only groups – underlining the importance of early public engagement in decision-making and planning processes.

We also found that there was less contention amongst the public when technological options (e.g. electric or hydrogen vehicles) were presented than when transport demand policy options were presented but that neither technological nor policy innovation was viewed as unproblematic. More barriers to behaviour change and acceptability were cited in respect of modal shift or demand reduction (e.g. lack of integrated public transport, absence of cycling infrastructure and unfairness of push measures; cf. King et al., 2009) than of adoption of electric, hydrogen or ultra-lightweight vehicles. Nevertheless, new technology vehicles, notably electric and hydrogen vehicles, were seen as too expensive and not ready for widespread use.

This finding is consistent with previous research on consumer responses to new transport technologies. Research on UK public attitudes to EVs suggests that understanding about EVs is limited, but interest in low-emission vehicles is high (e.g. DEFRA, 2009) demonstrated by the rapid growth in sales of small, energy-efficient vehicles (Nykvist and Whitmarsh, 2008). EST (2010) found two-thirds of the UK public would like a low-carbon car if they could afford one; 75% would consider fuel efficiency an important factor when buying their next car; but half of respondents do not know whether they can use an EV where they live and only one in four would consider an electric car at that time. In addition, few people thought EVs could now perform as well as conventional cars for many types of travel, and EV sales remain low in the United Kingdom (Vaughan, 2011). Similarly, Graham-Rowe et al. (2012) found that mainstream car consumers perceived the current generation of EVs as a 'work in progress' and too costly, despite offering environmental benefits. Other research shows cost savings, followed by environmental benefits, are the most important factors that would positively influence consumers' decision to purchase an EV as their next vehicle; access to charging points and battery driving range, and price, are the most important factors discouraging purchase of an EV (Ernst and Young, 2010). In some cases, members of the public have some (limited) experience with EVs (Graham-Rowe et al., 2012), but commonly public perceptions are based on little knowledge of current EV technologies. Indeed, initial findings from a European project on EV users (eBRIDGE, 2013) suggests that a majority of drivers found EVs better than they expected before driving them for the first time. This suggests a major barrier to EV uptake is familiarity and experience of EV use, as well as limited supporting infrastructure (cf. Geels, 2005).

4.4 Conclusions

In this chapter, we have highlighted the complex interplay of factors shaping transport-related attitudes and behaviour, and the challenges associated with changing these behaviours. In particular, we have called into question assumptions that vehicle choice and travel behaviours are usually a product of rational deliberation or that economic factors are most important in shaping consumer choices or their relationships with their cars. Rather, while transport demand has indeed risen broadly in line with income, modal choice is often shaped by unconscious habit and available infrastructure, and vehicle choice by social factors such as identity, status and interpersonal communication. Overall, the evidence we have reviewed paints a picture of car ownership and use as being intimately bound with cultural assumptions, physical infrastructure, ingrained practices and macro-trends in consumption. These factors contribute to a state of lock-in to personal automobility, even in cases where better transport options are available. Studies of radical innovation ('transitions') indicate that overcoming this lock-in will require multiple stakeholders (including the public) and levers, as well as processes of autonomous change, to shift the transport system towards a more sustainable direction (e.g. Geels, 2005; Nykvist and Whitmarsh, 2008). At the same time, more suitable roles for personal automobility (e.g. in rural areas) may not currently be optimally exploited. As noted in the introduction, we consider the public to operate within diverse decision-making contexts, not only consumption but also citizenship and as members of social groups (families, organizations, communities, etc.) and can thus influence transport systems and the automotive industry in multiple ways.

References

Anable, J. (2005). Complacent car addicts' or 'aspiring environmentalists'? Identifying travel behaviour segments using attitude theory. *Transport Policy*, **12**(1), 65–78.

Anable, J., Lane, B. & Kelay, T. (2006). *An evidence base review of public attitudes to climate change and transport behaviour*. Retrieved 1 September 2006 from http://www.dft.gov.uk/sustainable/climatechange. Accessed 25 March 2015.

Arrow, K. (2000). Increasing returns: Historiographic issues and path dependence. *European Journal of the History of Economic Thought*, **7**, 171–80.

Bakken, D. (2008). Car talk: The role and impact of word of mouth in brand choice. Presentation at ESOMAR Automotive Conference. Lausanne. http://www.warc.com. Accessed 25 March 2015.

Bamberg, S. (2006). Is a residential relocation a good opportunity to change people's travel behavior? Results from a theory-driven intervention study. *Environment and Behavior*, **38**, 820–840.

Barr, S. (2004). Are we all environmentalists now? Rhetoric and reality in environmental action. *Geoforum*, **35**, 231–249.

BBC News (British Broadcasting Corporation). Fuel crisis post mortem begins (13 September 2000). Retrieved 4 November 2013 from http://news.bbc.co.uk/1/hi/uk/925616.stm. Accessed 25 March 2015.

Black, C., Collins, A. & Snell, M. (2001). Encouraging walking: The case of journey-to-school trips in compact urban areas. *Urban Studies*, **38**(7), 1121–1141.

Blake, J. (1999). Overcoming the 'Value-Action Gap' in environmental policy: Tensions between national policy and local experience. *Local Environment*, **4**(3), 257–278.

Braun, O.L. & Wicklund, R.A. (1989). Psychological antecedents of conspicuous consumption. *Journal of Economic Psychology*, **10**, 161–187.

eBRIDGE (2013). *Empowering e-Fleets for Business and Private Purposes in Cities*. Technical Report. Project website: http://www.ebridge-project.eu/en/. Accessed 25 March 2015.

Choo, S. & Mokhtarian, P.L. (2004). What type of vehicle do people drive? The role of attitude and lifestyle in influencing vehicle type choice. *Transportation Research Part A* **38**, 201–222.

Davies, D.G., Halliday, M.E., Mayes, M. & Pocock, R.L. (1997). *Attitudes to Cycling: A Qualitative Study and Conceptual Framework*. Crowthorne, Berkshire: Transport Research Laboratory.

DECC (2010). *Energy Consumption in the UK: Data tables*. http://www.decc.gov.uk/en/content/cms/statistics/publications/ecuk/ecuk.aspx. Accessed 17 April 2015.

DEFRA (2002). *Survey of Public Attitudes to Quality Of Life and to the Environment: 2001*. London: DEFRA.

DEFRA (2009). *Attitudes and Behaviour Towards the Environment*. London: Department for Environment, Food and Rural Affairs.

Dietz, T. & Stern, P.C. (Eds.) (2008). *Public Participation in Environmental Assessment and Decision Making*. Washington, DC: National Academies Press.

Dittmar, H. (1992). *The Social Psychology of Possessions: To Have Is to Be*. New York: St Martin's Press.

Emmert, S., Van De Lindt, M. & Luiten, H. (2010). *BarEnergy: Barriers to changes in energy behaviour among end consumers and households*. Final Report. Amsterdam: The Netherlands Organisation for Applied Scientific Research – TNO.

Eriksson, L., Garvill, J. & Nordlund, A.M. (2008). Acceptability of single and combined transport policy measures: The importance of environmental and policy specific beliefs. *Transportation Research Part A*, **42**, 1117–1128.

Ernst & Young (2010). *Gauging interest for plug-in hybrid and electric vehicles in select markets Compared results*. Ernst & Young Global Automotive Center. http://www.ey.com/Publication/vwLUAssets/EY-Gauging_interest_for_plug-in_hybrid_and_electric_vehicles_in_select_markets/$FILE/EY-Gauging_interest_for_plug-in_hybrid_and_electric_vehicles_in_select_markets.pdf. Accessed 25 April 2015.

EST (2010). *Energy Saving Trust Attitude Tracker*. London: Energy Saving Trust.

European Commission (2011). *Transport 2050. Roadmap to a Single European Transport Area – Towards a competitive and resource efficient transport system*. 144 final http://ec.europa.eu/transport/strategies/2011_white_paper_en.htm. Accessed 10 October 2011.

Festinger, L. (1954). A theory of social comparison processes. *Human Relations*, **7**(2), 117–140.

Fujii, S., Garling, T. & Kitamura, R. (2001). Changes in drivers' perceptions and use of public transport during a freeway closure: Effects of temporary structural change on cooperation in a real-life social dilemma. *Environment and Behavior*, **33**(6), 796–808.

Geels, F. (2005). *Technological Transitions and System Innovations: A Co-Evolutionary and Socio-Technical Analysis*, Cheltenham: Edward Elgar.

Gerike, R. Gehlert, T., Richter, F. & Schmidt, W. (2008). Think globally, act locally – Reducing environmental impacts of transport. *European Transport\TrasportiEuropei*, **38**, 61–84.

Goodwin, P., Dargay, J. & Hanly, M. (2004). Elasticities of road traffic and fuel consumption with respect to price and income: a review. *Transport Reviews: A Translational Interdisciplinary Journal*, **24**, 275–292.

Graham-Rowe, E., Gardner, B., Abraham, C., Skippon, S., Dittmar, H., Hutchins, R. & Stannard, J. (2012). Mainstream consumers driving plug-in battery-electric and plug-in hybrid electric cars: A qualitative analysis of responses and evaluations. *Transportation Research Part A: Policy and Practice*, **46**(1), 140–153.

Harrabin, R. (5 October 2004). Fuel protest costs treasury £2bn yearly. *BBC News* (British Broadcasting Corporation). Retrieved 4 November 2013 from http://news.bbc.co.uk/1/hi/uk/3716346.stm. Accessed 25 March 2015.

Hetherington, P. & Ward, D. (11 September 2000). Fuel crisis looms as pickets hit depots. *The Guardian* (Guardian News and Media). Retrieved 4 November 2013 from http://www.theguardian.com/uk/2000/sep/11/oil.business. Accessed 25 March 2015.

Hetherington, P., Wintour, P. & Denny, C. (12 September 2000). Panic as oil blockade bites. *The Guardian* (Guardian News and Media). Retrieved 4 November 2013 from http://www.theguardian.com/uk/2000/sep/12/oil.business3. Accessed 25 March 2015.

Hiscock, R., Macintyre, S., Keams, A. & Ellaway, A. (2002). Means of transport and ontological security: Do cars provide psycho-social benefits to users? *Transportation Research Part D*, **7**, 119–135.

Jackson, T. (2005). *Motivating sustainable consumption: A review of evidence on consumer behaviour and behavioural change*. Guildford: Sustainable Development Research Network.

Jaensirisak, S., Wardman, M. & May, A.D. (2005). Explaining variations in public acceptability of road pricing schemes. *Journal of Transport Economics and Policy*, **39**(2), 127–154.

Kay, A. (2005). A critique of the use of path dependency in policy studies. *Public Administration*, **83**, 553–571.

King, S., Dyball, M., Webster, T., Sharpe, A., Worley, A., DeWitt, J., Marsden, G., Harwatt, H., Kimble, M. & Jopson, A. (2009) *Exploring public attitudes to climate change and travel choices: Deliberative research*. Department for Transport, London. http://www.sasig.org.uk/wp-content/uploads/2009/07/Attitudes-to-cl-ch-trnsprt_ITS-People PolicyScience_Jan-2009.pdf. Accessed 25 April 2015.

Köhler, J. (2006). Transport and the environment: The need for policy for long-term radical change: A literature review for the DTI FORESIGHT project on Intelligent Infrastructure Systems. *IEE Proceedings Intelligent Transport Systems*, **153**(4), 292–301.

Lally, P., van Jaarsveld, C., Potts, H. & Wardle, J. (2010). How habits are formed: Modelling habit formation in the real world. *European Journal of Social Psychology*, **40**(6), 998–1009.

Lane, B. (2009). Is the car scrappage scheme helping the environment? *The Guardian*. http://www.theguardian.com/environment/ethicallivingblog/2009/aug/21/car-scrappage-carbon-emissions. Accessed 25 April 2015.

Lethbridge, N. (2001). *Transport Trends: Understanding Attitudes to Transport Policy*. DETR. http://webarchive.nationalarchives.gov.uk/+/http://www.dft.gov.uk/pgr/statistics/datatablespublications/trends/2001articles/article2understandingattitud5436. Accessed 7 March 2010.

Lorenzoni, I., Nicholson-Cole, S. & Whitmarsh, L. (2007). Barriers perceived to engaging with climate change among the UK public and their policy implications. *Global Environmental Change*, **17**(3–4), 445–459.

Lyons, G., Goodwin, P., Hanly, M., Dudley, G., Chatterjee, K., Anable, J., Wiltshire, P. & Susilo, Y. (2008). *Public Attitudes to Transport: Knowledge Review of Existing Evidence*. Report for Department for Transport. Department for Transport, London. http://eprints.uwe.ac.uk/10352/. Accessed 25 April 2015.

Mannering, F. & Winston, C. (1991). Brand loyalty and the decline of American automobile firms. *Brookings Papers on Economic Activity: Microeconomics*, 1991, 67–114.

Maynard, M. (2007). Say 'Hybrid' and Many People Will Hear 'Prius'. *New York Times*. New York: The New York Times Company. Retrieved 7 November 2013 from http://www.nytimes.com/2007/07/04/business/04hybrid.html?ex=1188273600&en=6ca51be8ad9134cc&ei=5070. Accessed 25 March 2015.

McGeevor, K. (2009). Designing policy to influence consumers: Consumer behaviour relating to the purchasing of environmentally preferable goods. *A project under the Framework contract for economic analysis ENV.G.1/FRA/2006/0073 – 2nd*. Project Report. London: Policy Studies Institute. Retrieved 4 November 2013 from http://ec.europa.eu/environment/enveco/pdf/RealWorldConsumerBehaviour.pdf. Accessed 25 March 2015.

Mitchell, W.J., Borroni-Bird, C.E. & Burns, L.D. (2010). *Reinventing the automobile: Personal urban mobility for the 21st century*. London: The MIT press.

Moran, J. (2006). Crossing the road in Britain. *The Historical Journal*, **49**, 447–496.

Newman, D. (2013). Cars and consumption. *Capital and Class*, **37**(3), 454–473.

Niches (2012). *Guidelines for Implementers of Electric Cars in Car Share Clubs*. WG4 Report. http://www.niches-transport.org/fileadmin/NICHESplus/G4Is/21582_policynotesWG4_3.indd_low.pdf. Accessed 25 March 2015.

Nieuwenhuis, P. (2014). *Sustainable Automobility: Understanding the Car as a Natural System*. Cheltenham: Edward Elgar.

Nykvist, B. & Whitmarsh, L. (2008). A multi-level analysis of sustainable mobility transitions: Niche development in the UK and Sweden. *Technological Forecasting & Social Change*, **75**, 1373–1387.

O'Connor, R.E., Bord, R.J., Yarnal, B. & Wiefek, N. (2002). Who wants to reduce greenhouse gas emissions? *Social Science Quarterly*, **83**, 1–17.

Polzin, S.E., Chu, X. & McGuckin, N. (2011). *Exploring Changing Travel Trends*, presented at Using National Household Travel Survey Data for Transportation Decision Making: A Workshop, Transportation Research Board (http://www.trb.org); at http://onlinepubs.trb.org/onlinepubs/conferences/2011/NHTS1/Polzin2.pdf. Accessed 30 October 2013.

Richards, M.G. (2006). *Congestion Charging in London: The Policy and the Politics*. Basingstoke: Palgrave Macmillian.

Schiller, P.L., Bruun, E.C. & Kenworthy, J.R. (2010). *An introduction to sustainable transportation: Policy, planning and implementation*. London: Earthscan.

Schlag, B. & Teubel, U. (1997). Public acceptability of transport pricing. *IATSS Research*, **21**, 134–142.

Schuitema, G., Steg, L. & Rothengatter, J.A. (2010). The acceptability, personal outcome expectations, and expected effects of transport pricing policies. *Journal of Environmental Psychology*, **30**, 587–593.

Schuitema, G., Anable, J., Skippon, S. & Kinnear, N. (2013). The role of instrumental, hedonic and symbolic attributes in the intention to adopt electric vehicles. *Transportation Research Part A: Policy and Practice*, **48**, 39–49.

Scottish Government (2009). *SEABS '08: The Scottish Environmental Attitudes & Behaviours Survey 2008*. Edinburgh: The Scottish Government.

Siebenhüner, B. (2004). Social learning and sustainability science: Which role can stakeholder participation play? *International Journal of Sustainable Development*, **7**(2), 146–163.

Small, K.A. (2001). *Urban Transportation Economics*. London: Routledge.

Steg, L., Vlek, C. & Slotegraaf, G. (2001). Instrumental-reasoned and symbolic-affective motives for using a motor car. *Transportation Research Part F: Traffic Psychology and Behaviour*, **4**(3), 151–169.

Thornton, A., Evans, L., Bunt, K., Simon, A., King, S. & Webster, T. (2011). *Climate Change & Transport Choices Segmentation Model – A Framework for Reducing CO_2 Emissions from Personal Travel*. London: Department for Transport.

TNO (2006). *Review and Analysis of the Reduction Potential and Costs of Technological and Other Measures to Reduce CO_2-Emissions from Passenger Cars, Final report*. Delft: TNO.

Train, K.E. & Winston, C. (2007). Vehicle choice behaviour and the declining market share of U.S. automakers. *International Economic Review*, **48**, 1469–1496.

UKERC (2009). What policies are effective at reducing carbon emissions from surface passenger transport? http://www.ukerc.ac.uk/publications/what-policies-are-effective-at-reducing-carbon-emissions-from-surface-passenger-transport.html. Accessed 25 April 2015.

Urry, J. (1999). Automobility, *Car Culture and Weightless Travel: A Discussion Paper*. Lancaster: Lancaster University.

Vaughan, A. (2011). Electric car UK sales sputter out. *The Guardian*. 21 October. http://www.guardian.co.uk/environment/2011/oct/21/electric-car-uk-sales-sputter. Accessed 25 March 2015.

Veblen, T.B. (1899). *The Theory of the Leisure Class. An Economic Study of Institutions*. London: Macmillan Publishers.

Vergne, J.-P. & Durand, R. (2010). The missing link between the theory and empirics of path dependence: Conceptual clarification, testability issue, and methodological implications. *Journal of Management Studies*, **47**, 736–759.

Verplanken, B. & Wood, W. (2006). Interventions to break and create consumer habits. *Journal of Public Policy and Marketing*, **25**, 90–103.

Verplanken, B., Aarts, H. & van Knippenberg, A. (1997). Habit, information acquisition, and the process of making travel mode choices. *European Journal of Social Psychology*, **27**, 539–560.

Verplanken, B., Walker, I., Davis, A. & Jurasek, M. (2008). Context change and travel mode choice: Combining the habit discontinuity and self-activation hypotheses. *Journal of Environmental Psychology*, **28**, 121–127.

Walker, I. (2013). Vehicle characteristics as a predictor of perceived driver gender, age and personality. http://drianwalker.com/vehicles/badger.php. Accessed 7 July 2015.

Weber, E. (2010). What shapes perceptions of climate change? *Wiley Interdisciplinary Reviews (WIREs) Climate Change*, **1**, 332–342.

Whitmarsh, L. (2009). Behavioural responses to climate change: Asymmetry of intentions and impacts. *Journal of Environmental Psychology*, **29**, 13–23.
Whitmarsh, L., Swartling, Å. & Jäger, J. (2009). Participation of experts and non-experts in a sustainability assessment of mobility. *Environmental Policy & Governance*, **19**, 232–250.
Whitmarsh, L., Upham, P., Poortinga, W., Darnton, A., McLachlan, C, Devine-Wright, P. & Sherry-Brennan, F. (2011). Public Attitudes to Low-Carbon Energy – Research Synthesis. RCUK. http://psych.cf.ac.uk/home2/whitmarsh/Energy%20Synthesis%20FINAL%20%2824%20Jan%29.pdf. Accessed 1 May 2015.
Wynne, B. (1996). May the sheep safely graze? A reflexive view of the expert-lay knowledge divide. *Risk, Environment and Modernity: Towards a New Ecology* B. Szerszynski, S. Lash & B.E. Wynne, (Eds.), pp. 44–83. London: Sage.
Xenias, D. & Martineau, J. (2012). '*Getting there': Towards a community based initiative to change travel behaviour in a rural Welsh locality*. Project report. Fishguard and Goodwick, March 2012.
Xenias, D. & Whitmarsh, L. (2013). Dimensions and determinants of expert and public preferences for low-carbon transport policies and technologies. *Transportation Research Part A*, **48**, 75–85.

5

Car Manufacturing

Paul Nieuwenhuis
Centre for Automotive Industry Research and Electric Vehicle Centre of Excellence, Cardiff Business School, Cardiff University, Cardiff, Wales, UK

5.1 Background and Prehistory

In 1885, and within about 100 km of each other, Carl Benz and Gottlieb Daimler put an internal combustion engine in a vehicle and thereby initiated automobility and launched the car on its present trajectory. During the previous century, a number of technologies had converged to allow the development of the internal combustion-engined car. These technological innovations were fast-running internal combustion engines; bicycles, combining chain drive and a frame of steel tubes (the ability to make and bend tubes became crucial) and improved tyres; developments in horse-drawn vehicles in terms of chassis and body, suspension and so on.

The fact that we regard the innovations by Benz and Daimler as embodying the technologies of the first car is an important point because it highlights how we now see the car as primarily petrol-driven, despite the fact that steam-powered vehicles had been around since the late eighteenth century, while early battery-electric vehicles had been pioneered in the 1840s. Yet, it is only with the emergence of the petrol-driven car that we traditionally mark the birth of the car. More broadly, this period saw the adoption of two new power sources in society generally: petroleum and electricity, continuing a process of moving to more and more concentrated sources of energy over time (Huber and Mills, 2005). In the new automotive sector, however, the new petrol engines had to compete both with the established steam technology as well as the newly emerging use of electric power for machines. The development of fast-running industrial engines, suitable for powering petrol-powered vehicles, was the work of a few people, notably Otto in Germany, and the Belgian Lenoir. Soon after, Diesel, like Daimler and Benz based in Germany, though born in France, developed the compression-ignition engine, although its popularity had to wait a few more decades. The higher speeds of powered vehicles soon prompted improvements in suspension and steering, as well as tyres, the latter notably by Michelin in France, who developed the removable pneumatic tyre (Lottman, 2003).

The Global Automotive Industry, First Edition. Edited by Paul Nieuwenhuis and Peter Wells.

Cars started life as craft-made products in that they were made one-by-one by hand with each being different and each component being different as it was adapted to its neighbours in the subassembly. Very rapidly, major suppliers were set up, particularly in France, able to supply engines, gearboxes, axles and other key components, allowing standardization, a key precursor to mass production (Jeal, 2012). Jeal (2012) explains that standardization did in fact exist in Europe before Cadillac, Ford and others expanded on it in the United States. Such standardization was essential to allowing the production of the proprietary components that enabled the proliferation of early carmakers with very low levels of vertical integration. This process of standardization in Europe was in fact the primary enabler for this proliferation of the number of brands between about 1895 and 1905. The key building blocks of cars were now available to all. Many firms assembled cars from bought in components and limited themselves to fitting their own badge, itself often an outsourced part. The modular construction of cars at this time made this possible. Cars were almost invariably built up on a separate chassis frame, which carried all the components and subassemblies needed to make it move: engine, transmission, axles and wheels. This modular product is what Ford made at Highland Park, Michigan, the factory designed to build the iconic Model T (Nieuwenhuis and Wells, 1997, 2003, 2007). The body was a separate product and many car manufacturers sold their products as running chassis only. The owner would buy a body from a specialist coachbuilder, sometimes replacing it with a more modern one over time, a practice popular with the British royal family, for example (Smith, 1980).

5.2 Ford, Budd and Sloan: The History of Mass Car Production

The role of Ford in the history of mass production has become contested, partly as a result of our earlier work (Nieuwenhuis and Wells, 1997, 2003, 2007; Nieuwenhuis, 2014). In order to substantiate this analysis, it has to be understood that mass car production today is very different from the way the Ford Model T was built at Highland Park, Michigan. Ford's Highland Park facility often made kits of parts that were then sent to various near-market Ford facilities to be assembled locally and fitted with local bodies. This happened not only in various locations in the United States but also abroad in places such as Trafford Park in Manchester, various locations in Australia, or Cork in Ireland. By contrast, mass car making today is dominated by large, centralized, assembly facilities sourcing from world-wide networks of suppliers. The products of these large plants are dispatched in fully finished form to buyers worldwide via complex distribution networks largely featuring private dealerships (see Chapter 12). However, the main difference is that a modern car factory is centred around the production of body shells.

The notion that Ford was the first to mass produce cars is widely accepted. However, the car he mass produced, the Model T, was very much of its period, based on a modular approach to car making as used by the previous generation of craft builders: separate chassis and separate, wood-framed, coach-built or 'composite' body. This contrasts sharply with modern mass-produced cars, and the distinctions here are crucial; today cars use all-steel 'monocoque', 'unit' or 'unibody' construction, whereby a structural welded metal box or body-shell fulfils all structural functions. In other words, it acts as both body and chassis. This technology was made possible by the development by Edward Budd and his chief engineer Joe Ledwinka, of the all-steel welded body in Philadelphia between about 1910 and 1914, that is, at about the same time that Ford introduced the moving assembly line (Grayson, 1978; Nieuwenhuis and Wells, 1997, 2007). Thus, Nieuwenhuis and Wells (2007) argue that modern mass car manufacturing owes at least as much to Budd and Ledwinka as it does to Henry Ford.

Modern, Budd-type steel body technology involves considerable capital investments in press, press tooling, welding and painting technologies, but once these investments are made, they allow the low unit costs so typical of mass production. It is argued, therefore, that Budd's innovations constitute the very basis for the economics of the industry, notably in determining the crucial economies of scale in mass car production (Nieuwenhuis and Wells, 2003, 2007). This technological revolution involved the transition from the manufacture of modular cars from in-house components – typified by Ford's production system for the Model T at the Highland Park plant and later River Rouge in Michigan – to a situation where the manufacture of steel bodies formed the core activity. These were then assembled into cars from what today are largely outsourced components and sub-assemblies.

This system we have termed 'Buddism' delivered well for the automotive industry while it was enjoying steadily expanding consumer demand, although in the more mature motorized countries a degree of market saturation set in from about the 1970s onwards. This led to a situation whereby market forces began undermining the Budd paradigm – this is explored further from the market perspective in Chapter 12. Another development, emphasizing the fact that this change is of a 'socio-technical' nature, is that this new technology proved eminently suited to successive increases in labour productivity via automation, to which the Budd system is particularly well suited. This affected all three of the main stages of production (e.g. press, weld and paint) hence offering continuous cost reduction and quality improvements, while also leading to a steady decline in the number of people employed in car making (Andera, 2007; see also Chapters 6, 7 and 8).

We have now reached a situation whereby markets expect much shorter product cycles, as well as more visible differentiation, leading to more diverse product ranges (cf. Chapter 3). These pressures mean lower volumes per model and thus a real danger of losing economies of scale. This is one of the primary reasons why profitability of mass car making has shown a steady decline since its introduction in the 1920s. Attempts to recapture the economies of scale needed for profitable Budd-style car making are one of the principal reasons for the industry's efforts to globalize and consolidate. Like the economic system of which it has become an integral part, the automotive mass production system requires constant growth in volume and newly motorizing markets such as China, India or Brazil still allow large volumes of relatively standardized cars to be made and sold profitably thus extending the life of the traditional Ford–Budd mass production system. In fact, such markets provide the foundation of much of the current profitability of firms like Volkswagen (VW) and General Motors (GM) and as a result provided much of the impetus for the economic recovery from the 2008 recession in Europe and North America (see also Chapter 10).

An analysis of modern car manufacturing and the typical investments involved indicates that the principal areas of investment centre around the development and production of internal combustion engines, and in the manufacturing and painting of body shells. A typical automotive 'assembly' plant is thus primarily engaged in the pressing, welding and painting of car bodies, while engines are now made in dedicated facilities not always located near those assembly plants. The subsequent assembly of these bodies into finished cars by using largely bought-in components has become a secondary activity in terms of investment, albeit employing most of the labour involved. In fact, it can be argued that most of the components Ford spent considerable resources on making more efficiently and in higher volumes are now sourced from outside suppliers. Modern car manufacturers typically outsource between 60 and 80% of the ex-works value of a car. It could therefore be argued that Ford presided not only over the birth of mass car manufacturing, but also over the birth of the supply industry that it relies on.

The principal investments in an assembly plant, then, involve three processes: press shop (which includes the high cost of dedicated die-sets), body-in-white (where pressed panels are welded together – usually by robots) and paint (to protect the steel against corrosion, as well as meet market expectations regarding colour; see Andrews et al., 2006). Market expectations mean that car bodies are expected to change far more often than, for example, castings or powertrain components (i.e. engine and transmission), which are largely invisible to the customer or end user. It is therefore imperative for such body-related investments to be repeated regularly, for minor facelifts, as well as more significant sheet metal changes for the 'new' model to succeed the previous generation in a way meaningful – that is, visible to the market. This entails that even those elements that are not replaced are at least expected to be reconfigured, reprogrammed and updated.

The adoption of this all-steel body technology swept through the industry, especially in the United States, and by 1925, Budd all-steel technology already had a 50% share of US body production (Courtenay, 1987). With Ford's and Budd's innovations, mass car production was made possible. The final building block, however, involved the means to create a mass car market, to generate the necessary demand for the many more cars that could now be produced. This was the principal contribution of GM under Alfred Sloan, whose reign saw a number of key innovations, building on the introduction of large-scale vehicle finance through the foundation in 1919 of the General Motors Acceptance Corporation. During the 1920s, GM also introduced the trade-in as a down payment on a new car and the manufacturer and dealer-run used car business. Although trading-in was initially planned as a means of removing older vehicles off the road, it was soon realized that sales of used cars could become an additional profitable activity for dealers (Flink, 1988; Nieuwenhuis, 1994). GM also at this time developed the concept of a product range, allowing customers to gradually trade up from a basic Chevrolet, via Oakland (later changed to Pontiac and abolished as part of the response to the 2009 bail-out), Buick, Oldsmobile (another bail-out victim), La Salle (used briefly as an entry-level Cadillac sub-brand), to finally arrive at Cadillac on the principle that '…other cars are for getting you there; a Cadillac is for when you have arrived' (paraphrased from Dubois and Kremer, 2012). In addition, GM, in contrast to Ford in particular, began to focus much more on styling, colour – enhanced by paint supplier Dupont's majority shareholding in GM – and the 'look' of the car, while also promoting the then novel idea of planned obsolescence. This was achieved through the annual model change and strongly influenced other industrial sectors in the process, thereby creating the modern consumer culture (Flink, 1988).

5.3 Monocoque Construction: Budd's Impact on Car Design

The development of the monocoque, unitary body, unit body or unibody constitutes the next key phase in this transformation. The first Budd applications still used a conventional modular, separate body and chassis system, something that persisted in the United States for many decades in the form of the 'body-on-frame' system (Nieuwenhuis and Wells, 1997, 2007). Yet, the Budd company found already in the 1920s that by using single body-side stampings, some of the load from the chassis could be fed into the body, thus allowing a weight reduction for the car as a whole (Thum, 1928). Dodge Brothers were an early adopter of Budd all-steel technology from 1915 onwards and had close links with Budd. In 1923, it launched with Budd the all-steel Series 116-FOUR, the first Dodge all-steel closed body and one of the first in the United States. Dodge also pioneered baked enamel finish to replace its traditional paint and varnish system. These developments were made possible by the use of Budd technology (Hyde, 2005).

Budd also identified some potential problems of the unibody, notably the need for greater accuracy and the problem of assembling the car after the body has been built. This requires greater care on the part of the assembly workers as they move in and out of the body during the assembly process in order to avoid damage to the painted body. This is still an issue in modern car assembly, which is generally addressed by means of protective covers of the painted body during assembly and also by the removal of the doors after painting. The latter is a procedure pioneered by the Japanese, but it needs very careful production sequencing to ensure that the correct doors are reunited with the correct body and has therefore only been possible with the advent of more sophisticated manufacturing control systems.

Budd also expected a weight advantage from integrating body and chassis. Weight reduction was already more relevant in Europe than in the United States at this time. The United States with its own oil supplies was self-sufficient in automotive fuel, allowing a low fuel cost and low energy efficiency culture to become established among its car designers, builders and customers. While some European countries had oil reserves, even with low levels of motorization, none was self-sufficient in terms of vehicle fuel. For these reasons, European governments taxed vehicle fuel from an early stage, thus helping protect their balance of payments, as oil had to be imported; an even more pressing issue as the use of motor vehicles grew rapidly.

The combined system of Ford's mass production and Budd-style body technology, followed closely by Sloan's concept for building a mass car market gradually spread around the world. After the United States, Europe adopted mass production and Budd all-steel technology, the latter pioneered by Citroen, who made the bold move of taking the further step of moving to unitary construction in 1934. This then became the established model for mass-producing cars. Unit construction is still spreading with sport-utility vehicles (SUVs) rapidly moving from body-on-frame to monocoque construction.

5.4 Toyotism

Toyotism is the term often used for the widespread adoption of the Toyota Production System, also known as 'lean production' and the labour practices it introduced (see Chapters 6, 7 and 8). As mentioned earlier, Budd-style mass production of bodies proved very amenable to automation. Thus, we see over time a gradual but steady replacing of labour with capital. The car industry today produces more cars per person per year as a result of this. It also means that in most developed car making countries, fewer people are employed in car production than a few decades ago, despite higher production levels – productivity has thus improved as a result of capital investments (Andera, 2007).

The equipment needed to make engines, transmissions and, especially, bodies is very expensive. This proved a large hurdle and greatly increased entry costs to car manufacturing. After

World War II, Japan was occupied by the United States and Japanese carmakers had an opportunity to visit US car plants. While they were impressed with what they saw, they also realized that the cost of these systems was very high and thought about ways in which the equipment could be used more intensively, thus making it more affordable. This move, led by Toyota, looked at the Ford–Budd system and thought of ways of taking out what they perceived of as waste, or 'muda' thereby reducing cost. Toyota's novel approach became known as the Toyota Production System (TPS) and was popularized as 'lean production' (Womack et al., 1990).

Although initially kept within Toyota, the secrets of the TPS gradually spread as Toyota was forced to disseminate the system throughout its supply chain. Principles such as just-in-time and total quality management require the cooperation of the entire supply chain. This did mean that the TPS became more widely known and starting with Toyota's Japanese competitors, followed by the Americans, led by Ford, and from there the European car makers, gradually the whole car industry adopted lean production, or at least as many elements of it as they could manage. In the process, Toyota lost some of its competitive advantage.

It is important to understand automobility as a system, for changing some key aspects of the system can radically alter the system as a whole. This is the case, for example, if we change the core technologies outlined above: the all-steel body and the internal combustion engine, as explored in later chapters. For much of the twentieth century, the world car industry was dominated by US manufacturers and their global networks, notably Ford and GM, and to a lesser extent, Chrysler. There was no doubt that the American model in the automotive industry used to be dominant on the global stage. Recent crises at the once 'Big 3' (GM, Ford and Chrysler) combined with the collapse or near-collapse of many of their major suppliers have undermined this dominance and have led to a questioning of the business model that underpins it (Nieuwenhuis and Wells, 2009; Wells, 2010, 2013). The small number of players and large home market ensured that North America became one of the few locations where genuine economies of scale in car making could be achieved using the Ford–Budd system; it became the true home of mass production; however, as the market shares of the Detroit 3 are eroded by imports and foreign transplant operations in the United States, it becomes increasingly difficult to achieve these economies of scale.

5.5 Buddism in Crisis?

Despite the success of the Budd system throughout most of the twentieth century, in the established car markets, Buddist mass car making can now be said to be in crisis. Profitability has been declining for decades. The performance of GM illustrates this decline in profitability. In the late 1920s, GM's net profit margin approached 20%, yet by 2000 this had fallen to only 2.7% (Haglund, 2001). GM posted record losses and may have been technically bankrupt even before the recession of 2008–2009, but was, as the phrase goes simply 'too big to fail'. This marginal profitability is not atypical for the volume end of car making, although specialist producers such as BMW are still able to maximize their return through 'cost recovery', leveraging the value of their brands (Williams et al., 1994). BMW is even able to practice this approach on relatively high volumes, making it one of the most profitable firms in the industry with a return on sales typically in the 8–10% range.

It is clear that the present structure of the industry and its dominant business model are so closely linked with the adoption of Budd all-steel technology that any significant change will be possible only with the adoption of a different set of core technologies and an alternative business model by the automotive industry. Through adopting mass production, car manufacturing has gradually become more efficient. This is seen as a good thing. However, there is an increasing realization that this drive for efficiency may not make the system more resilient (Walker and Salt, 2006). Walker and Salt (2006) explain that, on the whole, humans favour efficiency, which seems preferable from various perspectives, such as simplicity, perceived more efficient use of resources, hence lower cost and so on. However, this same efficiency does make any system more vulnerable to change and external impact. They explain that although optimization is about efficiency, it tends to be '...applied to a narrow range of values and a particular set of interests, the result is major inefficiencies in the way we generate value for societies...' (ibid. 7). Being efficient in this sense leads to the elimination of redundancies. We tend to keep only that which we perceive to be directly and immediately useful. Yet, when the operating environment changes, other things may be more useful, and these may be the very things we have discarded in our quest for efficiency. It is for this reason that it leads to a steady loss of resilience (Nieuwenhuis and Lammgård, 2013; Nieuwenhuis, 2014).

At a basic level, this can be seen in the way in which the mass production system interfaces with the market. Here is an area of tension whereby the mass production system favours efficiency through a lack of diversity, while consumers, that is, markets favour diversity. The attraction of moving into new markets is that newly motorizing countries tend to demand automobility – virtually at any cost, but particularly they are willing to forego diversity in favour of standardized basic automobility, at least temporarily. As markets mature they demand and expect greater diversity – this tendency towards diversity being entirely natural in the broadest sense (Nieuwenhuis, 2014) – and as a result profits decline, as efficiency declines and costs rise. However, resilience may well increase as a result. During the 1990s, the academic proponents of 'lean' also became aware of this and many of them started to promote the concept of 'agility'. It came to be recognized that while some activities along a value chain were best done 'lean', in other areas, particularly those close to the market, agility was required, which led to the coining of the term 'leagile'.

5.6 Lean v Agile

Essentially, while lean was regarded as the answer for all supply chains at one stage (Womack et al., 1990), it was found that tensions could arise where some supply chains interface with markets. An overview of these developments was provided by Moyano-Fuentes and Sacristán-Díaz (2012). In view of some of the limitations of lean discovered in the literature by these authors, it was concluded that for certain supply chains and certain markets, a more agile approach was required, which at times could undermine the leanness of that supply chain. The question then became how lean and agile elements could be combined within a single supply chain and at what point in that supply chain the change from lean to agile should be made – this became known as the 'decoupling point'. Over time, two types of decoupling point came to be identified: the material decoupling point and the information decoupling point.

The 'decoupling point' concept was first mooted by Hoekstra and Romme (1992), and defined as 'the point in the product axis to which the customer's order penetrates. It is where order driven and forecast driven activities meet'. Later, Mason-Jones and Towill (1999) argued that there are at least two pipelines within the supply chain, material flow and information flow, and both flows have their own separate decoupling points. Therefore, they introduced the concept of 'material decoupling point' and 'information decoupling point'. This 'material decoupling point' resonates with the 'decoupling point' proposed by Hoekstra and Romme (1992). Mason-Jones and Towill (1999) defined the 'information decoupling point' as 'the point in the information pipeline to which the marketplace order data penetrates without modification. It is where market driven and forecast driven information flows meet'.

The understanding is then that upstream from the decoupling point, processes are operated on lean principles – inventory is held in generic form and the final configuration is only performed when the customer order is received – while downstream from the decoupling point, the agile principle is applied. Christopher and Towill (2001) stated that this approach can be applied when there is a possibility of modular design in product architecture. This resonates with aspects of the automotive value chain whereby generic subassemblies such as powertrain can be made by large facilities in order to achieve the desired economies of scale via lean processes. These are then supplied to assembly plants, which combine them in a variety of more flexible processes.

Several factors impact on the position of the material decoupling point. On the one hand, it depends on the longest lead time the end customer is prepared to tolerate (Mason-Jones and Towill, 1999; Naylor et al., 1999; Childerhouse and Towill, 2000; Mason-Jones et al., 2000). On the other hand, its position depends on the product variety and variability in demand. An increase in product variety and fluctuating volume of demand would force the material decoupling point to move upstream, which makes the supply chain more agile. In contrast, a more stable business environment with lower product variety and stable demand would move the material decoupling point downstream, making the supply chain leaner (Krishnamurthy and Yauch, 2007).

Naylor et al. (1999) argued that a 'postponement' strategy contributes to moving the material decoupling point closer to the end customer, thereby increasing both the efficiency and responsiveness of the supply chain. Postponement here refers to delayed configuration; the final assembly does not take place until customer orders are received (Christopher and Towill, 2000). Similarly, Childerhouse and Towill (2000) define postponement as the application of the material decoupling point before the point of product differentiation. A core element behind the postponement strategy is modular design. Feitzinger and Lee (1997) proposed two concepts: 'modular product design' and 'modular process design'. Modular product design refers to dividing the entire product into several sub-modules, and redesigning modules with standardized interfaces so that sub-modules can be easily assembled together, which enables components to be manufactured separately and even in parallel and one component can be shared by different products. Similarly, modular process design refers to breaking down the complete production process into several simple independent sub-processes that can function together as a whole, thus, the production sub-processes can be performed separately or can be re-sequenced. On the same basis, some processes can be performed in-house, while others can be outsourced, with this mix changing over time. Modularity enables a company to assemble standard components in the earlier stages of production and delay assembling the components that differentiate the products.

Postponement strategies contribute to leanness as well as agility. On the one hand, by delaying product differentiation, the supply chain produces standard semi-finished products as long as possible. Product differentiation then occurs at the material decoupling point, the generic inventory is regarded as strategic stock. Only differentiated processes cause delay, which greatly reduces the lead time from the order placement by customers to product delivery. It increases the responsiveness of the supply chain. This thinking has been behind the drive towards modular supply in the automotive industry, whereby suppliers no longer supply components, but comprehensive modules that are designed to interface with each other in the assembly process. Smart's facility in Hambach is probably one of the best examples of this approach.

However, little is actually said in the literature about the precise nature of the material decoupling point; it is often, by implication, suggested that greater freedom exists in reconfiguring supply chains than is reflected in reality. It is frequently implied that the supply chain designer has complete freedom in locating decoupling points. To the extent that it can be manipulated for strategic or tactical purposes, what determines the location of the material decoupling point in a supply chain, and what are therefore the constraints faced by supply chain designers and managers in managing this point? The analysis of the Budd system outlined above suggests that these constraints are often determined by levels of fixed investment in the equipment used at various processing stages along the supply chain. Where such fixed investments are high, lean is best, where they are low, agile works well. To an extent, this resonates with the notion of modularity in processes, whereby some processes are more amenable to agility, while others are of necessity lean (Feitzinger and Lee, 1997). Thus, in modern automotive manufacturing, to the extent this is feasible, those high capital investment processes associated with the all-steel body tend to be lean, while the low capital, high labour content processes of final assembly allow the agility needed to interface flexibly with the market.

5.7 Conclusions

In the mass production car industry, the key areas of production focus are the integrated steel body structure and the powertrain (engine and transmission). These major subassemblies represent the highest level of investment both in terms of product development and in capital investment in manufacturing. It is therefore these areas that represent both the key to economies of scale in the car industry and also the main barriers to greater agility in terms of response to customer requirements (Nieuwenhuis and Wells, 1997, 2003; Wells, 2010). It is for this reason that while customers often have access to a broad range of colour and trim variations, for example, their choice in terms of powertrain and body style is restricted, despite the now widespread adoption of platform strategies, modular supply and concerted attempts at 'mass customization' (Alford et al., 2000; Doran et al., 2007; Brabazon et al., 2010). It is also typically the case in mass car production that both the body and powertrain areas remain within the realm of the final assembler, the OEM (original equipment manufacturer; in the car industry this term is used for the final assembler, the firm whose badge appears on the car). For these reasons, much of Toyota's innovative work was focused on the area of body production – press shop, body-in-white and paint, as these are the least amenable to flexibility (Womack et al., 1990).

However, one way of breaking through this barrier to greater agility may be by moving to a more modular design approach, as suggested in the decoupling point literature (Feitzinger and Lee, 1997), and then moving the decoupling point for these key assemblies further up the supply chain, for example by outsourcing engines, and/or outsourcing body/chassis. Another method may be to move to a different type of technology that allows breakeven at much lower economies of scale, such as by abandoning 'Budd-style' all-steel body construction for a different solution (Nieuwenhuis and Wells, 2003, 2007; Nieuwenhuis, 2014). Some of these notions will be explored in later chapters.

References

Alford, D., Sackett, P. and Nelder, G. (2000), Mass-customisation – an automotive perspective. *International Journal of Production Economics* **65**, 99–11.

Andera, J. (2007), *Driving Under the Influence: Strategic Trade Policy and Market Integration in the European Car Industry, Lund Studies in Economic History*, **42**, Stockholm: Almqvist & Wiksell.

Andrews, D., Nieuwenhuis, P. and Ewing, P. (2006), Living systems, total design and the evolution of the automobile: the significance and application of holistic design methods in automotive design, manufacture and operation. *Design and Nature* **2**, 381–446.

Brabazon, P., MacCarthy, B., Woodcock, A. and Hawkins, R. (2010), Mass customization in the automotive industry: comparing interdealer trading and reconfiguration flexibilities in order fulfillment. *Production and Operations Management* **19**(5), 489–502.

Childerhouse, P. and Towill, D. (2000), Engineering supply chains to match customer requirements. *Logistics Information Management* **13**(6), 337–345.

Christopher, M. and Towill, D.R. (2000), Supply chain migration from lean and functional to agile and customized. *Supply Chain Management: An International Journal* **5**(4), 206–213.

Christopher, M. and Towill, D.R. (2001), An integrated model for the design of agile supply chains. *International Journal of Physical Distribution & Logistics Management* **31**(4), 235–246.

Courtenay, V. (1987), *Ideas That Move America...The Budd Company at 75*, Troy, MI: The Budd Company.

Doran, D., Hill, A., Hwang, K.-S. and Jacob, G. (2007), Supply chain modularization: cases from the French automobile industry. *International Journal of Production Economics* **106**, 2–11.

Dubois, C. and Kremer, E. (2012), *Cadillac*, New York: Assouline.

Feitzinger, E. and Lee, H.L.(1997), Mass customization at Hewlett-Packard: the power of postponement. *Harvard Business Review* **75**(1), 116–121.

Flink, J. (1988), *The Automobile Age*, Cambridge, MA: MIT Press.

Grayson, S. (1978), The all-steel world of Edward Budd. *Automobile Quarterly* **XVI**(4), 352–367.

Haglund, R. (2001), Tell-all tome: GM-published history includes company's woes, book review reprinted from *The Flint Journal in Society of Automotive Historians Journal* **191**, March–April, 8.

Hoekstra, S. and Romme, J. (1992), *Integrated Logistics Structures: Developing Customer Oriented Goods Flow*. London: McGraw-Hill.

Huber, P. and Mills, M. (2005), *The Bottomless Well: The Twilight of Fuel, the Virtue of Waste and Why We Will Never Run Out of Energy*, New York: Basic Books.

Hyde, C. (2005), *The Dodge Brothers: The Men, the Motor Cars and the Legacy*, Detroit: Wayne State University Press.

Jeal, M. (2012), Mass confusion: the beginnings of the volume-production of motorcars. *Automotive History Review* **54**, Autumn, 34–47.

Krishnamurthy, R. and Yauch, C.A. (2007), Leagile manufacturing: a proposed corporate infrastructure. *International Journal of Operations & Production Management* **27**(6), 588–604.

Lottman, H. (2003), *The Michelin Men: Driving an Empire*, London: Tauris.

Mason-Jones, R. and Towill, D.R. (1999), Using the information decoupling point to improve supply chain performance. *International Journal of Logistics Management* **10**(2), 13–26.

Mason-Jones, R., Naylor, B. and Towill, D. (2000), Engineering the leagile supply chain. *International Journal of Agile Management Systems* **2**(1), 54–61.

Moyano-Fuentes, J. and Sacristán-Díaz, M. (2012), Learning on lean: a review of thinking and research. *International Journal of Operations & Production Management* **32**(5), 551–582.

Naylor, J.B., Naim, M. and Berry, D. (1999), Leagility: integrating the lean and agile manufacturing paradigms in the total supply chain. *International Journal of Production Economics* **62**(1–2), 107–118.

Nieuwenhuis, P. (1994), The long-life car: investigating a motor industry heresy. in Nieuwenhuis, P. & P. Wells (eds), *Motor Vehicles in the Environment*, Chichester: John Wiley & Sons, Ltd.

Nieuwenhuis, P. (2014), *Sustainable Automobility: Understanding the Car as a Natural System*, Cheltenham: Edward Elgar.

Nieuwenhuis, P. and Lammgård, C. (2013), Industrial ecology as an ecological model for business: diversity and firm survival. *Progress in Industrial Ecology – An International Journal* **8**(3), 189–204.

Nieuwenhuis P. and Wells, P. (1997), *The Death of Motoring?: Car Making and Automobility in the 21st Century*, Chichester: John Wiley & Sons, Ltd.

Nieuwenhuis, P. and Wells, P. (2003), *The Automotive Industry and the Environment: A Technical, Business and Social Future*, Cambridge: Woodhead.

Nieuwenhuis, P. and Wells, P. (2007), The all-steel body as a cornerstone to the foundations of the mass production car industry. *Industrial and Corporate Change* **16**(2), 183–211.

Nieuwenhuis, P. and Wells, P. (2009), *Car Futures: Rethinking the Automotive Industry Beyond the American Model*, www.trendtracker.co.uk. Accessed 25 March 2015.

Smith, B. (1980), *The Daimler Tradition*, Isleworth: Transport Bookman.

Thum, E. (1928), Many advantages realized in body of 5-piece all-steel design. *Automotive Industries*, September **15**, 370–372.

Walker, B. and Salt, D. (2006), *Resilience Thinking: Sustaining Ecosystems and People in a Changing World*, Washington, DC: Island Press.

Wells, P. (2010), *The Automotive Industry in an Era of Eco-Austerity*, Cheltenham: Edward Elgar.

Wells, P. (2013), *Business Models for Sustainability*, Cheltenham: Edward Elgar.

Williams, K., Haslam, C., Johal, S. and Williams, J. (1994), *Cars, Analysis, History, Cases*, Providence, RI: Berghahn.

Womack, J., Jones, D. and Roos, D. (1990), *The Machine That Changed the World*, New York: Rawson.

6

Recent Trends in Manufacturing Innovation Policy for the Automotive Sector: A Survey of the United States, Mexico, European Union, Germany and Spain

Patrick Galvin, Elena Goracinova and David A. Wolfe
Innovation Policy Lab, Munk School of Global Affairs, University of Toronto, Toronto, Ontario, Canada

6.1 Introduction

The global economy continues to experience a rapid and dynamic period of transformation with manufacturing activities being decomposed and recomposed on a global basis.[1] Within this perspective, manufacturing is often viewed as a low value-added activity that is increasingly performed in developing economies. A growing body of literature, however, questions this view by focusing on novel trends in manufacturing and the critical importance of this sector for innovation across the broader economy (Sirkin et al., 2011; Manyika et al., 2012). The application of information technology and more sophisticated advanced manufacturing technologies makes the locational choices of large multinational firms and domestic ones infinitely more complex than it has been in previous decades. While labour costs were long considered to be the driving force in these locational decisions, there is recognition that regional assets – such as skills training, research and development and a strong supply base of

[1] Research support for this chapter was provided by Automotive Partnership Canada and the Social Sciences and Humanities Research Council of Canada under Grant No. 886-2012-0001, which is gratefully acknowledged.

The Global Automotive Industry, First Edition. Edited by Paul Nieuwenhuis and Peter Wells.

technologically sophisticated manufacturers – are increasingly important (Christopherson et al., 2010). There is also a growing acknowledgement that the loss of manufacturing activity eventually undermines the capacity of advanced economies to retain their research and production base (Berger, 2013). This has led to a renewed interest in industrial policy that aims to strengthen manufacturing competitiveness.

In recognition of these trends, leading economies, such as the United States, the European Union (EU), Germany and Spain, are all moving towards establishing a twenty-first century manufacturing innovation policy designed to strengthen their respective automotive industries. Recent initiatives adopted in these jurisdictions share a long-term focus on the continued development and diffusion of general-purpose technologies, such as new materials, IT and batteries, which are transforming and greening the automotive industry (Mazzucato, 2013; Perez, 2013; Tassey, 2014). In this sense, the reinvigorated automotive policy combines the horizontal basis for generating enabling technologies across a wide swath of manufacturing industries, with a more sectoral application on the automotive industry. These initiatives support not only basic knowledge generation but also the subsequent competitive technology production, aiming to support new productive capabilities across broad technology platforms with potential benefit for a cross-section of manufacturing industries. In the context of this emerging policy approach, industrial policy instruments aim to address both market and systems failure (Crafts and Hughes, 2013). Automotive-related manufacturing policies have sought to address systems failures stemming from the difficulty of transitioning from the dominant internal combustion engine (ICE) paradigm to low carbon vehicles and the increasing incorporation of electronics in automobiles.

There is a commitment to the promotion of cross-disciplinary and inter-sectoral collaboration in response to the increasingly interdisciplinary nature of technological development and the reliance of automotive assemblers on partners providing supporting technologies and services. In other words, firm-centric policy supports are being displaced by a new policy approach focused on promoting linkages between firms, research institutions and government to stimulate platform-based and network-oriented innovations (O'Sullivan et al., 2013). Key policy tools are cluster initiatives and alternate forms of public–private partnerships that bring together cross-sectoral knowledge and expertise to 'help build systems, create networks, develop institutions and align strategic priorities' (Warwick, 2013, 4; Wessner, 2014). Other measures being adopted, which feed into manufacturing innovation policies, provide support for small- and medium-sized enterprises (SMEs), the supply of labour, finance and increased company access and presence on international markets. For multifaceted manufacturing innovation programmes to succeed, officials aim to improve policy coordination across multiple levels of governance and strategically important areas, such as basic and applied research, standardization measures, as well as deployment and market access (European Commission, 2013).

Despite the common view that US political institutions are less hospitable to industrial policy than those of their competitors in Europe and East Asia, network forms of organization are gaining greater prominence as mechanisms of governance and emerging as precisely the kind of innovation-focused mechanism the US policy is targeting. Non-market forms of coordination have traditionally been more common in the German style of industrial organization, but are becoming more decentralized as Germany seeks to retain its leadership in high-growth markets. Spain has also developed innovative strategies to secure cooperation, despite the lack of strong non-market coordination mechanisms. In the case of Mexico, policy-makers have mainly focused on the attraction of foreign direct investment initially through the use of special automotive decrees, but then increasingly over the past two decades through the negotiation of preferential trade agreements (PTAs) with North America, the EU and Japan (Manger, 2009). Notwithstanding this emphasis, there are emergent efforts to facilitate greater regional cooperation for economic development through the support of clusters at different levels of governance (Carillo, 2009).

The purpose of this chapter is to outline the contributing factors that are changing the nature of manufacturing and their impact on the automotive industry. Secondly, it provides a general description of the essential features of current manufacturing and automotive-related policies in the United States, Mexico, the EU, Germany and Spain. It concludes with a reflection on the prominent role of state policies in the ongoing shifts in the automotive industry towards a new technological paradigm.

6.2 A Changing Manufacturing Landscape

Firms today are faced with shorter product life cycles, hard-to-manufacture product designs and increasing technological parity, which require an exceedingly diverse range of capabilities (Dicken, 2007). This leads a growing number of firms to focus on 'core competences' and rely on foreign and domestic manufacturing and non-manufacturing firms for everything else in order to save time and resources and enhance flexibility (Herrigel, 2010; Thun, 2010). However, the aggressive offloading of low value-added manufacturing operations to developing countries has begun to hollow out supply chains in industrial economies. Outmigration has expanded to include higher value-added production operations and services, bringing into question the dichotomous view of economic activities in terms of the production of goods and services (Andreoni and Gomez, 2012). At the core of these developments has been the inability to completely separate R&D/design and manufacturing processes because of complementary relationships between product and process innovation (Pisano and Shih, 2009; Breznitz and Cowhey, 2012). Another factor has been the gradual development of the national innovation systems (the organized collection of public and private assets that create and utilize technologies) in emerging economies, which can provide the environment where companies can transfer both low and high value-added operations (Tassey, 2010). These developments compromise the ability of advanced economies to capture economic value and generate employment because product evolution often occurs elsewhere. Even advanced economies with substantial support for manufacturers, such as Germany, are facing rising competition from emerging economies.

Lastly, developments in key enabling technologies are rapidly changing the competitive basis for manufacturing industries (Weyer et al., 1997). A case in point is the transformation of the automobile from a 'modestly complex set of hardware components into today's modern automobile, which contains 17 subsystems for which electronics is a central element' (Tassey, 2014, 29–30). Furthermore, the dwindling of natural resources has increased demand for product and manufacturing energy and resource efficiency. Tassey (2010) underlines the relevance of public funding for enabling technologies since the technical and market risks associated with their development lead firms to under-invest in the platform and infratechnologies required for successful commercialization.

6.3 Restructuring in the Automotive Industry

Current automotive production is strongly geographically concentrated with North America, Europe and East Asia making up 90% of the total production (Dicken, 2013). Developments in the past four decades have led to the spectacular growth of the Japanese and Chinese automobile industries, while the US market share has fallen dramatically. Dicken (2013) also notes the significant growth of automobile production in Spain, Mexico, South Korea, Brazil, India, Russia and the emerging market economies of Eastern Europe. Furthermore, while the 2008–2009 financial crisis hurt automotive production in developed economies, output grew in developing countries such as Brazil, China and India. The consumption side is characterized by similar trends, with slowing demand in the mature European and North American markets, while Asia is expected to be the source of major growth.

These changes in the geography of the automotive industry have coincided with the transition from the mass production model of manufacturing automobiles to an age of lean production, based on the modularization of certain components, the development of component systems and standardized platforms (see Chapter 5). This shift has led to the current era of intensified competition, in which the dominant market position of the leading producers has been challenged by rising competitors from some of the BRIC countries. The pressures associated with intensified competition and overcapacity among established producers have led major assemblers to devolve as many functions as possible to a complex, tiered network of suppliers (Berger, 2005, 58). The production facilities of suppliers in lower wage regions are no longer limited to performing low value-added operations, but have progressed up the value chain in a new form of practice that blends R&D/design and manufacture capabilities across global production locations. In both North America and Germany, the outsourcing of production to abroad has contributed to an outflow of automotive R&D (Jürgens and Krzywdzinski, 2009; Klier and Rubenstein, 2014). Unlike the US case, however, German domestic production has been maintained despite the growing production capacity abroad. This ongoing pressure on suppliers to participate in product development, while reducing costs and improving quality, has increased consolidation as a way for firms to keep up with customer demands (Andreoni and Gomez, 2012). Finally, the move towards electric mobility is confronting assemblers and suppliers with another period of radical change along the automotive sector's entire value chain. The integration of electric components, advanced materials, batteries and new high-performance electronic architectures into their products make automotive manufacturers increasingly reliant on suppliers from a cross-section of other industries, which is resulting in a blurring of the sectoral boundaries.

Despite the decentralization of production, distance continues to matter in the automotive industry (Schmitt and Biesebroeck, 2011). A production location in proximity to an assembly plant increases the chances that a supplier is awarded an outsourcing contract because it diminishes transportation costs, product development time and helps avoid costly misunderstandings (Winden, 2011). This factor contributes to a common trend – particularly prevalent in North America and Europe, but also to a certain extent in East Asia – the continued regionalization of production and distribution in the automotive industry (see also Chapter 15). Despite these developments, Dicken (2013) argues that production geographies and supplier–customer relationships are influenced by the national context and specific histories. Major European automobile producers are strongly Eurocentric in comparison to the more heavily transnationalized character of US automotive Original Equipment Manufacturers (OEMs). Berger (2005) notes that in comparison to their German or Japanese counterparts, American companies hand over more of the responsibility for module design and assembly to their tier one suppliers. These trends in both the technological basis of automotive production and the organization of the industry over the past four decades have led to major policy shifts among the leading automotive producing nations. The following section outlines the differences and similarities in policy responses to the ongoing transformation of the automotive industry.

6.4 Automotive Policies in the United States, Mexico, EU, Germany and Spain

6.4.1 United States

In recent decades, the United States placed less emphasis on the relevance of manufacturing in comparison to its German counterparts (Pisano and Shih, 2012). Moreover, the commercialization stage of innovation was largely seen as the purview of the private sector. Strong market regulation was not complemented by a coherent national industrial policy, which has made the initiation and coordination of collaborative efforts across agencies and government levels difficult to achieve (Bonvillian, 2012). This weak institutional basis contributes to the ad hoc basis of numerous networking initiatives. A study conducted by the MIT Production in the Innovation Economy (PIE) Project concludes that US local ecosystems differ from industrial districts in Northern Italy or Southern Germany, where producers draw on the close ties they develop with actors in their local ecosystems such as suppliers, customers or regional innovation centres (Locke and Wellhausen, 2013).

The 2008 crisis prompted a reevaluation of the industrial structure of the US automotive industry and revealed the structural weakness in the sector. The recession inflicted significant damage on the US auto industry. This decline in employment, sales and production was especially severe for the so-called Detroit Three carmakers: combined US sales for Chrysler, Ford and GM dropped from a level of 8.4 million in 2007 to 6.4 million in 2008 and 4.6 million in 2009. In addition, their combined market share fell from 52% in 2007 to 48% and 44% in the next 2 years (Klier and Rubenstein, 2013). As the recession deepened, the financial condition of both Chrysler and GM grew more precarious, so much so, that by December 2008 both companies could not secure the credit they needed to conduct their daily operations (Cooney et al., 2009).

When the Obama administration took office the following year, it established a Presidential Task Force on the auto industry on 16 February 2009. Based on its investigation of the state of the US auto industry, the Task Force came to the conclusion that the only chance of survival for Chrysler and GM lay in bankruptcy that would allow them to restructure. However, in an unconventional move, the Task Force made use of an established, but less frequently adopted, part of the bankruptcy code, Section 363, to achieve the restructurings. Under this section, a newly formed company could buy any desired assets from the bankrupt firm and immediately begin operating as a solvent corporation (White House, 2009). The Task Force was also fortunate to be operating under TARP rules put in place the previous November that allowed it to allocate capital flexibly, without having to seek Congressional approval. The availability of TARP financing provided the auto firms with access to the financing needed to restructure and allowed them to avoid the more dramatic alternative of liquidation. As a result of the restructuring efforts, the companies were able to begin operating as private enterprises once again as the Task Force had anticipated (Rattner, 2009).

In the aftermath of this dramatic restructuring of two of the leading US automotive firms, the federal government launched a set of initiatives targeted at the manufacturing sector more broadly, but also organized around a number of national priorities. They encourage closer cross-agency as well as industry–university–government coordination (Bonvillian, 2012; PCAST, 2011). Automotive sector-specific programmes follow on from Department of Energy (DoE)-coordinated government–industry automotive R&D initiatives launched during the Clinton and Bush administrations, or the Partnership for a New Generation of Vehicles (PNGV) and FreedomCAR respectively (Sperling, 2003). The Energy Independence and Security Act passed in 2007 authorized the Advanced Technology Vehicle Manufacturing (ATVM) programme that provided $25 billion in direct loans to automobile and component manufacturers to support projects intended to promote fuel efficiency and reduce the overall dependence in the US economy on petroleum (Wessner, 2012, 75). Whereas the PNGV and FreedomCar initiatives promoted the advancement of hydrogen fuel and fuel cell vehicles, a significant portion of ATVM funds support the development of advanced batteries and electric drive components for transportation electrification. It was followed by a 2009 government programme, which allocated $2 billion in grants to 30 factories that are producing advanced batteries and electric drive components. In 2011, the DoE also announced the US Drive cooperative partnership, which in addition to government expertise draws on the know-how of industry partners from the automotive, energy and electric utility industries. The partnership has the mandate to advance innovation and commercialization in a range of automotive and energy infrastructure technologies.[2]

Related initiatives aim to promote manufacturing and energy-efficient technologies, which are key to the modernization of the automotive industry and the revitalization of regional economies. Leading US R&D agencies, including the Departments of Defense and Energy, have intensified their efforts in the advanced manufacturing space. The 2013 fiscal year budget called for $500 million in funding for the DoE to aid advanced manufacturing in flexible electronics and lightweight vehicles, another $200 million to be allocated to Defence Advanced Research Projects Agency (DARPA) for research to support advanced manufacturing and further increases in funding to the National Science Foundation for programmes in cyber physical systems, robotics and advanced manufacturing (Wessner, 2012, 76). These initiatives were complemented by the establishment of the Advanced Research Projects Agency for energy, known as ARPA-E intended to speed up the commercialization of clean-energy technologies (Weiss, 2014).

Another essential initiative is the National Network for Manufacturing Innovation, as part of which regional Institutes for Manufacturing Innovation will attempt to accelerate the development and adoption of advanced manufacturing technologies. The network was originally announced in March 2012 as a nation-wide strategy to boost local and state economies and national competitiveness by organizing regional public–private university partnerships focused on technology development in specific areas. The Obama administration now hopes to continue to expand this programme into a nationwide network of 45 Institutes over the next decade, according to a White House fact sheet. President Obama called on Congress to pass legislation to double the currently planned number of institutes through legislation that has already been introduced in both houses.[3]

---

[2] http://www.uscar.org/guest/partnership/1/us-drive

[3] http://ssti.org/blog/president-urges-expansion-national-manufacturing-network-state-union

While the federal government has been the key institutional actor in the new US policy landscape, state governments have also adopted measures to support the future of the automotive sector. As the home of the North American auto industry, Michigan has proposed measures to support research into advanced automotive technologies. At the outset of the crisis, Michigan's Economic Development Corporation launched the Michigan Battery Initiative which succeeded in attracting $1.3 billion in ARRA funds to support the initiative, in addition to the over $1 billion in grants, tax credits and skills-related programmes offered by the state to support manufacturing of lithium–ion battery cells, packs and components. State and federal support for the sector generated considerable private sector activity with 16 projects underway by 2010. However, by 2012, one of the leading producers had been bought out by a Chinese firm and much of the sector still faced considerable challenges from East Asian producers. Despite these headwinds, the initiative is noteworthy for the state-led focus it represents (Wessner, 2013, 75–83). South Carolina (SC) offers another example of a state engaged in a major effort to upgrade the technological capabilities of its innovation clusters, especially in the automotive industry. The South Carolina Research Authority has the mandate to build innovation systems to generate more commercial results from the research output of the state's three universities – Clemson University, the University of South Carolina and the Medical University of SC. One of the most notable results has been the 2003 establishment of the Clemson University International Center for Automotive Research (CU-ICAR). Over the past decade, the centre has grown to include a 90 000 ft^2 facility with world-class faculty holding fully endowed chairs. The centre has proven attractive to private firms in the state including a $25 million contribution from Bayerische Motoren Werke AG (BMW) for the construction of the graduate engineering centre of CU-ICAR (Wessner, 2012, 458). Overall, non-market coordination – often associated with Coordinated Market Economies – has been an increasingly prominent component of twenty-first century US industrial policy, building on and extending previous federal government collaborative initiatives, such as the Manufacturing Extension Partnership Program.

6.4.2 Mexico

The policy instruments that played a critical role in the growth of the Mexican non-indigenous automotive sector differ significantly from the other country case studies. The period of state-led industrialization was followed by an export-led strategy driven by foreign direct investment in 'maquiladora' assembly operations based on imported inputs. Policy-makers attracted foreign direct investment initially through the use of special automotive decrees, but then increasingly over the past two decades through the negotiation of PTAs with North America, the EU and Japan (Manger, 2009). Mexico has reconfirmed its steadfast dedication to foreign investment through the 2007 establishment of ProMexico, by way of Presidential Decree, as a sectoral public trust controlled by the Ministry of the Economy to promote FDI, along with the export of goods and services. Most of the FDI is used to construct new vehicle and engine export plants, as well as to increase the capacity of existing plants (Ministry of Economy, Government of Mexico, 2012). Human capital availability and efficient logistics in the central region automotive cluster have contributed to the proliferation of higher value activities such as engineering and design in addition to low-skills part manufacturing (Miroff, 2013). The push for research activities, however, has been dominated by foreign-owned OEMs, which limits decision power for generating and transferring technology and demonstrates the difficulty of the periphery achieving core status. Import competition driven by OEMs has also decimated the number of small- and medium-sized firms that produced parts for Mexico's original equipment and parts markets (Carillo, 1998).

Notwithstanding these challenges, both the national and municipal governments have begun to establish intermediary organizations that attempt to capture the synergies between the automotive industry, the knowledge infrastructure and regional governments as the basis of regional economic development. The regional focus is a product of the industry's concentration into four large regional clusters in the North, Central and Bajio areas. A prominent competitiveness initiative, called SINTONIA, was launched in the state of Puebla in the Central Region automotive cluster. SINTONIA's management structure is composed of representatives from the main clusters in the state along with government representatives, societal organizations leaders, universities, directors and researchers. It is coordinated by an interdisciplinary group of academics and researchers who have expertise in innovation and competitiveness projects. The universities and research institutes that are involved in this project include the Universidad Popular Autónoma del Estado de Puebla, the Institute for Strategy and Competitiveness, as well as Harvard Business School. The ultimate goal and intention of SINTONIA is to be a 'neutral broker providing the thought leadership on clusters and the infrastructure needed to convene all the stakeholders' (Nuño de la Parra and Quitl, 2013, 833).

6.4.3 European Union

In the post-2008 crisis period, the EU has renewed its interest in industrial policy, with a particular emphasis on the automotive industry. The CARS 2020 Action Plan provided a roadmap for automotive-related spending under the Europe 2020 strategy, which deploys funds through instruments such as the Research Framework Programs and the Cohesion Policy (European Commission, 2012). CARS 2020 builds on four main pillars: (i) investing in advanced technologies and financing innovation, (ii) improving market conditions, (iii) enhancing global competitiveness and (iv) investing in human capital and skills and softening the social impacts of restructuring.

Major Research Framework Programs that have tackled the goals espoused in CARS 2020 include the European Green Cars Initiative (EGC) and its successor, the European Green Vehicles Initiatives (EGVI). They take a non-territorial approach and often function as public–private partnerships that encourage cross-country collaborative research projects to advance and integrate automotive, energy, Information and communications technology (ICT) and smart grid technologies. The consultative process in both the EGC and EGVI builds on the expert advice of three industry-led European Technology Platforms-ERTRAC, EPoSS and SmartGrids. Furthermore, an International Non-profit Association (EGVIA) was set up for the purpose of facilitating lasting cooperation between partners. Most EU-funded research on hydrogen and fuel cells is channelled through the Fuel Cells and Hydrogen Joint Technology Initiative (JTI). The founding members of this venture are the European Community and a non-profit association of European industry interests composed of a major share of Europe's fuel cell and hydrogen firms of all sizes from micro to larger transnational corporations (European Commission, 2010). It is another essential public–private cross-sectoral partnership initially launched in 2008 and renewed in 2013, with around a third of the 2008 JTI's budget devoted to developing hydrogen-fuelled road vehicles and a corresponding hydrogen refuelling infrastructure, as well as other supporting elements needed for a wider uptake of hydrogen-powered vehicles.

Over the period from 2007 to 2013 around 8 billion Euros were invested in activities linked to research, development and innovation, including helping domestic and cross-border networks develop research-intensive products and services. They include investment in key enabling technologies, such as automotive electronics and hydrogen and fuel cell technology leading to the transformation of regional economies around new cross-sectoral domains, such as electro-mobility. Finally, a significant portion of funding support is directed towards SMEs. In addition to grants received from the Research Framework and Cohesion policy initiatives, funding is delivered through new financial instruments, including loan schemes (Horizon 2020 Financial Instruments, 2014). Current loan volumes under the research framework programme include EUR 11.3 billion for more than 90 projects in 23 countries (EC, 2013). Any organization that is involved in transport research can apply for loans from the European Investment Bank. These are the best financial instruments to use for a project that involves greater risk. The two programmes that are most often used for this type of research are the (i) the Risk-Sharing Finance Facility (RSFF) and (ii) the European Clean Transport Facility (ECTF) loan instruments. Lastly, another source of support are demand-side and public procurement measures such as the reduction of circulation and registration taxes for low CO_2 cars (Government of Aragón, Ministry for Science and Technology, 2009, 13). Along with CO_2 emissions regulation, EU initiatives in support of private–public co-development of general-purpose technologies could lead to the greening of the automotive industry.

6.4.4 Germany

Germany's automotive industry relies on a vast network of publicly funded intermediaries such as the Fraunhofer Institutes and AiF (German Federation of Industrial Research Associations) for its applied research needs. Concerns, however, have arisen over the automotive industry's ability to modernize, considering the dominance of large corporate actors in research networks and the resulting technological lock-in in the ICE technological paradigm (Kleinert, 2012). Despite its status as an important player in environmental policies, the relevance and power of the automotive sector have also created strong resistance to limiting CO_2 emissions from cars. Notwithstanding, policies encompassed in the recently introduced High-Tech Strategy (HTS) have the potential to address the technological lock-in into the ICE technology in the automotive industry. Schneider and Weyer (2012) argue that the HTS extends research policy to an integrated and coordinated industrial policy geared towards competitiveness and innovation. It is committed to increasing subsidies to research and demonstration projects aligned with missions such as mobility and climate change, which are particularly relevant to the automotive industry. Novel initiatives are also shaped by ongoing processes of regionalization prompted by EU integration and attempts to reform German federalism by increasing the scope for subnational autonomous political action.

Following the guiding principles incorporated into the current round of EU cohesion and regional development policies (McCann and Ortega-Argilès, 2013), German federal and state governments have adopted a smart specialization approach and promoted regional public–private initiatives including clusters to strengthen multilevel networks within and across sectors (Kiese, 2008; PRBS, 2012). Automotive-related projects are also part of federal-level comprehensive strategies to advance green growth and catch up to competitors including the

National Electromobility Development Plan and the 6th Energy Research Program. In other words, the federal government develops macro-frameworks for the promotion of specific missions. Simultaneously, there are state-level thematic strategies under which public officials mobilize and coordinate regional actors to establish cross-sectoral demonstration and R&D projects as well as compete for federal and EU funds. Consequently, local governments in Germany (as in many other Western European countries) have become increasingly involved in developing networks and new relationships with business interests. There has also been an emphasis on promoting science–industry collaboration across the board, which has brought about relatively novel roles for German universities governed by the Länder. Even though German federalism is described as consensus driven, these relatively novel collaborative initiatives stimulate interregional competition and make intergovernmental cooperation more conflict ridden (Eickelpasch and Fritsch, 2005; Staehler et al., 2007). Despite the large R&D effort, production subsidies and consumer incentives for EVs have been limited. This reflects a generally less interventionist position of German governments. Germany's stance on financial incentives might be subject to change since the National Electromobility Development Plan has continuously underlined their relevance for the 2014–2017 market ramp-up phase especially considering the goal of introducing one million electric vehicles (EVs) by 2020.

6.4.5 Spain

The Spanish government has not shied away from intervention in the automotive industry. In comparison to other peripheral regions, Spain has gradually experienced changes in its institutional environment conducive to the presence of high value-added automotive operations. The key elements of the auto industry development in Spain are the nationally and regionally funded automotive clusters in various regions aimed at promoting cooperation, project development and joint ventures across actors from related sectors. They have a strong OEM presence and a developed network of automotive suppliers strengthened through supplier association with SERNAUTO (Spanish Association of Automotive Equipment and Components Manufacturers). More recently, SERNAUTO became the coordinator of the Spanish Automotive Technology Platform SERtec, which was created in 2005 and launched in March 2006 with government support. Its objective is to promote R&D cooperation between all stakeholders involved in the development of vehicles, with particular emphasis on enabling automotive suppliers overcome technological challenges.

In the aftermath of the 2008 crisis, however, the sector is facing serious challenges. Industrial policies to ameliorate the economic downturn have had a strong territorial and sectoral orientation as well as an emphasis on facilitating science–industry–public sector collaboration, despite the lack of strong non-market coordination mechanisms. In 2009, Spain introduced the comprehensive Plan for the Automotive Industry that supports automotive manufacturers both through a scrappage programme and a Competitiveness Plan aimed at the modernization and adoption of advanced production systems, the industrialization of new products and processes, including Green Vehicles, as well as the training of new staff. The government has invested more than 4 billion Euros to enable the industrial maintenance, sustainability and employment of this sector in the medium term (Spanish Public Employment Service, 2012, 27).

Within the commitment of the Spanish government to promote EVs, the MOVELE programme (Electric Mobility Pilot Project) merits special mention. The MOVELE programme was an electromobility pilot project that ran from 2009 to 2010 and was defined in the framework of the Action Plan for Energy Efficiency and Saving (2008–2011). The goal of the project was to introduce 2000 EVs into urban environments while also building at least 500 recharging spots so that these vehicles could be recharged. Following MOVELE's success, Spain's national government also presented the 2010 'Integral Plan for the Promotion of Electric Vehicles', which includes an 'Integrated Strategy for EVs 2010–2014' promotion in Spain that established a target of having one million H&EVs on the road by the end of 2014 (Government of Spain, Ministry of Industry, Tourism, and Trade, 2010, 14).

Lastly, in contrast to its US and German counterparts, but in tandem with the EU, the Spanish government advanced an explicit industrial policy or the Integrated Industrial Policy 2020 (PIN 2020) in late 2010. The goal is that over the next 10 years, the Spanish industrial policy should combine a 'horizontal orientation, seeking higher competitiveness of the Spanish industrial companies, with sector focused actions… based on their growth potential, as well as on the need for the modernization and redefinition of their mature industries' (Government of Spain, Ministry of Industry, Tourism, and Commerce, 2009, 2). The strategy identifies the automotive industry as key to the Spanish economy and forecasts a substantial economic impact on the automobile sector in the time period between 2011 and 2015. The PIN 2020 seeks to position the Spanish automotive industry as one of the top automotive industries in the world, and it plans to do this by investing for the future in the most competitive market segments such as hybrid, electric and reduced emission vehicles. Overall, Spain has continued to build an innovation ecosystem conducive to the development of novel technologies and their diffusion in the automotive industry (Government of Spain, Ministry of Industry, Tourism, and Commerce, 2009, 2).

6.5 Conclusion

Recent policy developments reaffirm the potential of specific forms of publicly supported innovative activity, often the product of policy learning across national jurisdictions, to help facilitate the modernization of the automotive industry. Leading automotive producing nations are imposing more stringent emissions standards, while simultaneously engaging in efforts to support the transition of their respective automotive industries to a new, more environmentally sustainable technological paradigm. Automotive policies are designed to address the increasing internationalization and agglomeration of economic activities, as well as the development, diffusion and commercialization of radical new enabling technologies. Particular emphasis has been placed on the commercialization stage of innovation as countries seek to capitalize on domestic innovative activity to drive a green industrial revolution in manufacturing. Within this context, the internal innovation capacities of automotive firms are being complemented by government-supported strategic R&D alliances as well as new forms of inter-sectoral and interdisciplinary cooperation between public and private research institutions. There are also programmes to provide increased support for SMEs, which lack the internal resources to engage in the development and adoption of transformative technologies.

Given the unique historical institutional and economic trajectories of individual countries or jurisdictions, there are shared and distinct challenges as well as approaches to modernizing the automotive industry. Focusing on the similarities, Germany, Spain and the United States are engaging in systematic efforts to enhance coordination and create more innovation-based manufacturing platforms in new and emerging technologies across a range of actors. Mexico has taken a more trade-focused approach that seeks to capitalize on its lower labour costs and export platforms, but has also begun to establish public–private partnerships that build on regional assets in attempts to boost industry competitiveness. In each case study, however, the complex and networked character of recent public policies and firm organization reflects emergent and hybrid institutional arrangements. In conclusion, this account of the evolving role of governments in automotive policy demonstrates the prominence of new forms of state engagement in domestic processes of technological transformation.

References

Andreoni, A., and C. L. Gomez. 2012. Can We Live on Services? Exploring Manufacturing-services Interfaces and Their Implications for Industrial Policy Design. Proceedings of DRUID Academy 2012, University of Cambridge/ The Moeller Centre, 2012.

Berger, S. 2005. *How We Compete: What Companies Around the World Are Doing to Make It in Today's Global Economy*. New York: Currency Doubleday.

Berger, S. 2013. *Making in America: From Innovation to Market*. Cambridge, MA: MIT Press.

Bonvillian, W. B. 2012. Reinventing American Manufacturing: The Role of Innovation. *Innovations* **7**: 3.

Bonvillian, W. 2014. The New Model Innovation Agencies: An Overview. *Science and Public Policy* **41**(4): 425–37.

Breznitz, D., and P. Cowhey. 2012. *America's Two Systems of Innovation: Recommendations for Policy Changes to Support Innovation, Production and Job Creation*. San Diego: Connect Innovation Institute.

Carillo, V. J. 1998. Productivity, Income and Labor in the Automotive Industry in Mexico. In *Incomes and Productivity in North America: Papers of the 1998 Seminar*. Dallas, TX: Commission for Labor Cooperation, North American Agreement on Labor Cooperation.

Carillo, J. 2009. Les generations d'entreprises maquiladoras: Une analyse critique. *Cahiers des Amerique latine* (56): 27–43.

Christopherson, S., J. Michie, and P. Tyler. 2010. Regional Resilience: Theoretical and Empirical Perspectives. *Cambridge Journal of Regions, Economy and Society* **3**(1): 3–10.

Cooney, S., Bickley, J.M., Chaikind, H., Petit, C.A., Purcell, P., Rapaport, C., and Shorter, G. 2009. U.S. Motor Vehicle Industry: Federal Financial Assistance and Restructuring. (Report No. R40003). Washington, DC: Congressional Research Service. Retrieved from http://www.fas.org/sgp/crs/mise/R40003.pdf. Accessed 22 April 2015.

Crafts, N., and A. Hughes. 2013. Industrial Policy for the Medium to Long-Term. Working paper. Cambridge: Cambridge University.

Dicken, P. 2007. *Global Shift: Mapping the Changing Contours of the World Economy*, 5th ed. London: SAGE Publications.

Dicken, P. 2013. *Global Shift: Mapping the Changing Contours of the World Economy*. 6th ed. New York: SAGE Publications.

Eickelpasch, A., and M. Fritsch. 2005. Contests for Cooperation – A New Approach in German Innovation Policy. *Research Policy* **34**(8): 1269–282.

European Commission. 2010. An Integrated Industrial Policy for the Globalization Era: Putting Competitiveness and Sustainability at Centre Stage. A Communication from the Commission to the European Parliament, The Council, and The European Economic and Social Committee, and The Committee of the Regions, Brussels.

European Commission. 2012. EC rolls out CARS 2020 action plan for European Auto Industry. Available at: http://www.greencarcongress.com/2012/11/cars2020-20121108.html. Accessed 25 March 2015.

European Commission. 2013. A European Strategy for Key Enabling Technologies – A Bridge to Growth and Jobs. Communication from the Commission to the European Parliament, the Council, the European Economic and Social Committee and the Committee of the Regions. European Commission. Web.

Government of Aragon, Ministry for Science and Technology. 2009. Spanish Capabilities in the Eco-electro Road Mobility Sector and the FP7 Green Cars Initiative. Tudela, Spain: Government of Aragon.

Government of Spain, Ministry of Industry, Tourism, and Commerce. 2009. Integral Plan of Industrial Policy, PIN 2020.

Government of Spain, Ministry of Industry, Tourism, and Trade. 2010. Spanish initiatives in the Electromobility Sector. Workshop on Research for the Fully Electric Vehicle, Brussels, 11 June 2010.

Government of Spain, Spanish Public Employment Service. 2012. Prospective Study of the Automotive Sector in Spain. Published by the Subdirectorate-General for Statistics and Information Organization and Planning Management Area Occupational Observatory in Madrid.

Herrigel, G. 2010. *Manufacturing Possibilities: Creative Action and Industrial Recomposition in the United States, Germany, and Japan*. Oxford: Oxford University Press.

Horizon 2020 Financial Instruments. 2014. Presentation. European Commission, 17 October 2014. Web. 30 December 2014.

Jurgens, U., and M. Krzywdzinski. 2009. Changing East-West Division of Labour in the European Automotive Industry. *European Urban and Regional Studies* **16**(1): 27–42.

Kiese, M. 2008. Cluster Approaches to Local Economic Development: Conceptual Remarks and Case Studies from Lower Saxony, Germany. In: B., Uwe, and M., Gunther, eds., *The Economics of Regional Clusters: Networks, Technology and Policy*. New Horizons in Regional Science. Cheltenham: Edward Elgar. 269–303.

Kleinert, I. 2012. Ready for Change? Electric Vehicle Technologies Challenge the German Automotive Innovation System. Gerpisa Colloquium, 2012, Krakow.

Klier, T., and J. M. Rubenstein. 2013. Restructuring of the U.S. Auto Industry in the 2008–2009 Recession. *Economic Development Quarterly* published online 18 March 2013. Available at: http://edq.sagepub.com/content/early/2013/03/14/089/1242413481243. Accessed 25 March 2015.

Klier, T., and J. Rubenstein. 2014. The Evolving Geography of the U.S. Motor Vehicle Industry. In: F. Giarratani, G. Hewings, and P. McCann, eds., *Handbook of Industry Studies and Economic Geography*. Cheltenham: Edward Elgar. 38–66.

Locke, R. M., and R. L. Wellhausen. 2013. *Production in the Innovation Economy*. Cambridge, MA: MIT Press.

Manger, M. 2009. *Investing in Protection: The Politics of Preferential Trade Agreements between North and South*. New York/Cambridge: Cambridge University Press.

Manyika, J., J. Sinclair, R. Dobbs, G. Strube, L. Rassey, J. Mischke, J. Remes, C. Roxburgh, K. George, D. O'Halloran, and S. Ramaswamy. 2012. *Manufacturing the Future: The Next Era of Global Growth and Innovation*. Washington, DC: McKinsey Global Institute.

Mazzucato, M. 2013. *The Entrepreneurial State*. London: Anthem Press.

McCann, P., and R. Ortega-Argilès. 2013. Modern Regional Innovation Policy. *Cambridge Journal of Regions, Economy and Society* **6**(2): 187–216.

Ministry of Economy, Government of Mexico. 2012. Automotive Industry Monograph. Department of Heavy and High Technology Industries, Del. Cuauhtemoc Mexico, Distrito Federal, March 2012.

Miroff, N. 2013. With Mexican Auto Manufacturing Boom, New Worries. This Washington Post Blog. Available at: http://www.articles.washingtonpost.com2013-07-01/world/40294191.... Posted on 1 July 2013.

Nuño de la Parra, J. P., and E. V. Quitl. 2013. SINTONIA: A Model of Collaboration and Innovation for Puebla and the Region. *US-China Education Review B* **3** (11): 828–34. Web. 8 November 2014.

O'Sullivan, E., A. Andreoni, C. Lopez-Gomez, and M. Gregory. 2013. What Is New in the New Industrial Policy? A Manufacturing Systems Perspective. *Oxford Review of Economic Policy* **29**(2): 432–62.

Perez, C. 2013. Unleashing a Golden Age after the Financial Collapse: Drawing Lessons from History. *Environmental Innovation and Societal Transitions* **6**: 9–23.

Pisano, G. P., and W. C. Shih. 2009. Restoring American Competitiveness. *Harvard Business Review* **87**(July–August)114–25.

Pisano, G. P., and W. C. Shih. 2012. *Producing Prosperity: Why America Needs a Manufacturing Renaissance*. Boston: Harvard Business Review Press.

PRBS. 2012. Lebensqualität und Arbeitsplätze: Allianz für die Region GmbH treibtab 2013 die Regionalentwicklung voran. Press Release, 19 November 2012. http://www.projekt-region-braunschweig.de. Accessed 28 November 2012.

President's Council of Advisors on Science and Technology (PCAST). 2011. Report to the President on Ensuring American Leadership in Advanced Manufacturing, June, 2011, Executive Office of the President, Washington, DC.

Rattner, S. 2009. The Auto Bailout: How We Did It. Available at: http://www.money.cnn.com/2009/10/21/autos/auto_bailout_rattner.fortune. Accessed 25 March 2015.

Schmitt, A., and J. Van Bisebroeck. 2011. Does Proximity (still) Influence the Choice of External Suppliers?. Web.

Sirkin, H. R., M. Zinser, and D. Hoehner. 2011. *Made in America Again, Why Manufacturing Will Return to the US*. Boston: Boston Consulting Group.

Sperling, D. 2003. *FreedomCAR and Fuel Cells: Toward the Hydrogen Economy?* Policy Report. Berkeley: Progressive Policy Institute.

Staehler, T., D. Dohse, and P. Cooke. 2007. Evaluation der Fördermassnahmen BioRegio und BioProfile. Report commissioned by BMBF (Bundesministerium für Bildung und Forschung), Consulting für Innovations- und Regionalanalysen, Institut für Weltwirtschaft and der Universität Kiel, Centre for Advanced Studies in the Social Sciences.

Tassey, G. 2010. Rationales and Mechanisms for Revitalizing US Manufacturing R&D Strategies. *The Journal of Technology Transfer* **35**(3): 283–333.

Tassey, G. 2014. Competing in Advanced Manufacturing: The Need for Improved Growth Models and Policies. *The Journal of Economic Perspectives* **28**(1): 27–48.

Thun, E. 2010. The Globalization of Production. In: J. Ravenhill, ed., *Global Political Economy*, 2nd ed., Oxford: Oxford University Press.

Van Winden, W. 2011. *Manufacturing in the New Urban Economy*. London: Routledge.

Warwick, K. 2013. *Beyond Industrial Policy: Emerging Issues and New Trends*. OECD Science, Technology and Industry Policy Papers, No. **2**, Paris: OECD Publishing. http://dx.doi.org/10.1787/5k4869clw0xp-en.

Weiss, L. 2014. *America Inc.? Innovation and Enterprise in the National Security State*. London: Cornell University Press.

Wessner, C. W. 2012. *Rising to the Challenge: U.S. Innovation Policy for the Global Economy*. National Research Council. Washington, DC: National Academies Press.

Wessner, C. W. 2013. *Best Practices in State and Regional Innovation Initiatives: Competing in the 21st Century*. National Research Council. Washington, DC: National Academies Press.

Wessner, C.W. 2014. Growing Innovation Clusters for American Prosperity: Summary of a Symposium. Rep. Committee on Competing in the 21st Century: Best Practice in State and Regional Innovation Initiatives, n.d. Web. 8 November 2014.

Schneider, V., and J. Weyer. 2012. Power Games in Space: The German High Tech Strategy and European Space Policy. *Innovation Policy and Governance in High-tech Industries: The Complexity of Coordination*. Ed. J.M. Bauer, A. Lang, and V. Schneider. Berlin: Springer.

Weyer, J., U. Kirchner, L. Riedl, and J. F.K. Schmidt. 1997. *Technik, die Gesellschaft schafft – Soziale Netzwerke als Ort der Technikgenese*. Berlin: Edition Sigma.

White House. 2009. Obama Administration New Path to Viability for GM and Chrysler. Retrieved from: http://www.whitehouse.gov/assets/documents/Fact_Sheet_GM_Chrysler.pdf. Accessed 22 April 2015.

7

Labour Relations and Human Resource Management in the Automotive Industry: North American Perspectives

John Holmes
Department of Geography, Queen's University, Kingston, Ontario, Canada

7.1 Introduction

Methods of manufacturing and work organization developed in the automotive industry have achieved paradigmatic status and influenced the organization of production in many other sectors of the economy. The auto industry has also played a pivotal role in shaping the institutional arrangements governing national systems of labour relations in a number of countries (Kochan et al., 1997). This is certainly true in the case of the United States and Canada where the automotive industry played a leading role in shaping the broader post-war labour relations system.[1] This influence stemmed from the sheer scale and economic importance of the automotive sector and the innovative collective agreement provisions bargained between autoworker unions and the D-3 auto makers (General Motors (GM), Ford and Chrysler).[2] Provisions such as multi-year contracts with cost-of-living (COLA)

[1] Writers associated with the French Regulation School highlight this industrial relations system, and especially the wage relation which saw a linking of real wage increases to productivity increases, as a key institutional element of the mode of social regulation underpinning Fordism in the United States (Aglietta, 1979; Lipietz, 1987; Boyer, 1990).
[2] For many years, GM, Ford and Chrysler were referred to as the Big Three but, following the significant rise in North American-based production by companies such as Toyota and Honda, are now commonly referred to as the Detroit 3, or simply the D-3.

The Global Automotive Industry, First Edition. Edited by Paul Nieuwenhuis and Peter Wells.

escalators and built-in annual wage increases, supplementary unemployment benefits, '30 years and out pensions', quality of work life programmes and pattern bargaining became common features of the US and Canadian collective agreements across both private and public unionized sectors and also set the standard for many large non-union employers (Hunter and Katz, 2012: 1985).

Although developed from a series of nationally based industries, the automotive industry today is often regarded as the epitome of a truly global industry. Dicken (2011: 322), for example, writes that 'the intensely transnational nature of automobile production renders the concept of "national" automobile industries virtually meaningless, although governments like to believe otherwise'. The industry certainly is global in the sense that through foreign direct investment a relatively small number of global automakers have established vehicle assembly facilities in each and every major global regional market and integrated their production networks and component supply chains across national boundaries within those regional blocs and beyond (see Chapter 2). Similarly, the automotive components industry is dominated by a few score global companies, which follow and often co-locate close to their customers' assembly facilities. Both automakers and Tier 1 component manufacturers plan their technological, production and marketing operations at a global level. Sturgeon et al. (2009), however, suggest that the contemporary auto industry exhibits a complex and, in some ways, surprising geography. For example, Stanford (2010: 385) emphasizes that 'while international trade flows are important, the industry still demonstrates a significant regionalization of production and investment'. This is because automakers continue to prefer locating final assembly near to their various markets, forced on them in part to the persistence of different local regulations and expectations (see Chapter 2). Thus, around 80% of vehicles purchased in the North American Free Trade Agreement (NAFTA) bloc are assembled in North American facilities and highly integrated production networks span facilities located in the United States, Mexico and Canada.[3] So, what are the implications for employment relations of the specific form assumed by globalization in the automotive industry?

There are wide-ranging debates regarding the impact of globalization on employment relations in general (Katz and Darbishire, 2000; Lansbury et al., 2003; Smith et al., 2008; Edwards et al., 2007). Some argue that globalization erodes national policy autonomy and is leading to a dismantling of nationally specific institutions governing employment relations. This will result, they argue, in a convergence over time towards an international norm in which employment relations are shaped solely by market relations. Conversely, other writers underscore the continued resilience of national institutions which mediate the impact of international economic pressures on firms and workers in a specific national economy (Wailes et al., 2008). They point to numerous studies highlighting continuing differences in employment relations across countries as providing empirical support for their argument that globalization is not imposing a universal logic on employment relations.

[3] The integration of the auto industry across the United States, Canada and Mexico occurred in stages as trade was liberalized by the 1965 US–Canada Auto Pact and later by the 1994 NAFTA (Holmes, 1993; Anastakis, 2005, 2013). Today, the following automakers operate assembly plants in North America: GM; Ford; Chrysler; Toyota; Honda; Nissan; Subaru; Mazda; Hyundai; Kia; BMW; Daimler; VW; Audi.

Since competition in the auto industry is no longer confined to individual countries, auto industry employment relations in a particular country may well be shaped more by developments in the global auto industry than by developments in other sectors of the national economy (Kirsch and Wailes, 2012: 1968). The evolution of advanced manufacturing technologies – as outlined in Chapter 4 – the frequent redesign of production lines for model changeovers, and the transnational reach of automakers explored in Chapter 2, all contribute to increased standardization across the global industry of the 'hard' production technologies used to manufacture motor vehicles. At the same time, significant variations in 'soft' technologies – labour relations (LR), human resource management (HRM) practices and shop-floor work organization – continue to persist between countries. As global automakers' corporate work system templates and LR/HRM practices diffuse internationally, they must be 'translated' and modified to mesh with specific national and local systems of industrial relations, labour market institutions and work cultures (Sorge, 2005: 182). Thus, an important 2008 comparative study of globalization and employment relations in the auto assembly industry in seven countries (Australia, China, Germany, Japan, Korea, Sweden and the United States) concluded that:

> … despite strong convergence pressures associated with increased international competition, as well as the growing transnationalization of production and the emergence of standard production systems, national arrangements continue to play a significant role in shaping employment relations outcomes in this industry. (Lansbury et al., 2008: 143)

Given the difficulty generalizing across countries about industrial relations and human resource management practices, discussion in this chapter is restricted to the United States and Canada. Notwithstanding some differences in auto industry LR/HRM practices (Kumar and Holmes, 1996; Rutherford and Holmes, 2013), they are sufficiently similar to warrant joint treatment.[4] In both countries, the prevailing legislative frameworks governing labour relations and collective bargaining are similar in origin and substance and, for close to half-a-century, autoworkers in the two countries were represented by the same international union – the United Automobile, Aerospace and Agricultural Implement Workers of America (United Automobile Workers (UAW)). Although some divergence in collective bargaining strategies and outcomes occurred after Canadian autoworkers left the UAW in 1985 to form their own independent National Automobile, Aerospace, Transportation and General Workers Union of Canada (CAW) (Kumar and Holmes, 1997), industry restructuring over the last decade has narrowed such differences (Rutherford and Holmes, 2013).[5]

The composition of the industry in both countries has changed dramatically since the 1980s. Prior to that, the industry was dominated by the D-3. Assembly and captive parts plants were 100% unionized and union density was significant in the independent automotive parts sector. Starting in the mid-1980s, an increasing number of Japanese, and later Korean and European, automakers and many of their principal suppliers set up 'transplant' operations in the United States and Canada. The transplants introduced elements of the so-called Japanese Production

[4] Mexico is an increasingly important part of the integrated North American automotive industry but the history and institutional structure of labour relations in Mexico are very different to that in the United States and Canada.

[5] On 31 August 2013, the CAW joined with the Communications, Energy and Paperworkers Union of Canada (CEP) to form a new union: UNIFOR. For the purposes of this chapter, we will continue to refer to CAW.

System (JPS – see also Chapter 6) into the North American auto industry and, with the exception of the early Japanese joint-venture transplants, remain resolutely non-union.[6] Union density in the parts industry fell dramatically over the last two decades as a result of the closure of many D-3 unionized in-house parts plants, the influx of non-union transplants and the growth of non-union domestic component companies. Employment relations vary considerably between, on the one hand, the un ionized D-3 automakers and their traditional domestic suppliers and, on the other hand, the non-unionized North American subsidiaries of Asian and European automakers and component manufacturers (Katz, 2008). Furthermore, industry restructuring produced increased variation in employment practices between D-3 companies and across plants within each company (Hunter and Katz, 2012). A principal theme running through the discussion in the remainder of this chapter is one of continuity and change in North American auto industry labour relations.

7.2 From Fordist Production to Lean Production: The Evolution of Labour Relations/Human Resource Management Systems in the North American Auto Industry Prior to 2000

This section traces the evolution of the post-war labour relations system in the North American automotive industry prior to 2000. It begins with the development of the archetypal Fordist system in the 1930s and 1940s (or from a more technology-based approach, the Ford–Budd system, as outlined in Chapter 4), which produced a distinctive and highly uniform pattern of labour relations across the auto industry in the United States and Canada. Then, in the 1980s, Japanese automakers and their key suppliers established a production bridgehead in the United States and Canada and introduced key elements of Japanese production methods (JPS) to North America. Corporate management at the D-3 embraced the idea of the JPS as a new production paradigm capable of superior performance and sought to implement elements of the JPS in their own assembly plants. D-3 and the UAW began negotiating substantial changes to labour relations practices, especially in the United States, and, by 1990, had produced a hybridized form of production organization and labour relations in D-3 plants which came to be known as 'lean production' (Womack et al., 1990; see also Chapter 4).

7.2.1 The Classic Fordist Industrial Relations System in the US and Canadian Automotive Industries

A distinctive North American labour relations system emerged from the institutional restructuring of the 1930s and governed labour–management relations and collective bargaining in the US and Canadian automobile industry for close to half a century. Two events in 1935 were fundamental in the development of the Fordist automotive labour relations system. First was the passage of the National Labor Relations Act (NLRA), commonly referred to as the Wagner Act, which provided the legal and administrative framework for contemporary

[6] The unionized joint-venture US transplants were established in the 1980s by Toyota, Mazda and Mitsubishi with GM (NUMMI closed in 2010), Ford (AutoAlliance) and Chrysler (Diamond Star) respectively. In Canada, workers at the GM-Suzuki joint-venture (CAMI now solely GM) are represented by the CAW. To date, union attempts to organize workers in the other US and Canadian transplants have failed.

American labour law.[7] The NLRA institutionalized collective bargaining and facilitated the development of a new model for union organizing – the industrial union. This legal and administrative structure was actively inscribed into the economic landscape through the political struggles and clashes between capital and labour, and between different factions within the labour movement.[8]

To organize mass production workers, who were largely unskilled, industrial unionism adopted a structure which organized workers not, as with craft unionism, by skill or occupation but by workplace. Once organized, all production workers employed in a particular plant belonged to a specific legally defined 'local' bargaining unit of a single national union. This geographically decentralized organizational structure could well have resulted in localized and fragmented collective bargaining. However, in a number of industries, but especially in the auto industry, it was offset by the development of an institutionalized system of pattern and connective bargaining linking national and local plant level bargaining: a system that produced uniformity in wages and working conditions across the industry (Katz, 1985; Holmes, 1991).

The second development was the formation of the UAW – one of the new industrial unions – which rapidly organized virtually all US and Canadian workers employed in the D-3s assembly and 'captive' in-house component plants and a significant number of workers in the independent automotive parts sector. The landmark 1948 GM-UAW collective agreements covered operations in both the United States and Canada and established a pattern that endured for the next 30 years.

These collective agreements were structured around three key elements: compensation determined in multi-year national collective agreements by formula-like wage rules which implicitly linked annual real wage increases to productivity increases; a connective bargaining structure linking national- and plant-level bargaining in which wages, benefits and general work rules were set in the national contract while local agreements defined plant-specific work rules such as job classifications and seniority ladders; and a job-control focus which linked worker rights to strictly defined job classifications and provided for the resolution of contractual disagreements between labour and management (Katz, 1985).[9] The connective bargaining system ensured uniformity in wages, benefits and general contract provisions across plants within one company and pattern bargaining ensured similar uniformity across the D-3 (Holmes, 1991; Katz et al., 2002).[10]

The Fordist labour relations system resulted in steadily rising real income for American and Canadian autoworkers enabling unskilled blue-collar workers to attain an affluent American

[7] Notwithstanding some important specific differences, the legal framework for the post-war industrial relations system in Canada closely paralleled that in the United States (Holmes and Rutherford, 2013).

[8] Struggles over the NLRA continued for at least a decade following its passage. The Taft–Hartley amendment (1947) represented a partial victory for employers in their effort to restrict the power granted to unions under the NLRA. Taft–Hartley prohibited secondary picketing and, most importantly, granted individual states the prerogative to enact anti-union right-to-work laws.

[9] Pay was determined by the wage rate attached to the specific job being performed rather than the characteristics of the worker undertaking it. The job classifications were grouped into career ladders with workers moving up the job ladder according to seniority and being laid-off in reverse order of seniority during downturns. Whilst this system facilitated the layoff of workers, it severely restricted management's ability to redeploy workers within a plant.

[10] Pattern bargaining worked by setting identical contract expiration date for each of the Big Three. As contract renewal time approached the union would select one target company for bargaining; the company from which it thought it could wrest the best contract. Once agreement had been reached with that company, with or without a strike, the other auto-makers quickly settled for virtually identical contracts.

middle-class life style: the American Dream. Katz et al. (2002: 71) noted that 'from the late 1940s until the late 1970s, the application of wage rules and job control unionism produced steadily rising real compensation to auto workers and long-term growth in auto employment and production. With limited import penetration in auto sales, this was a bargaining process where the geographic bounds of union organization closely matched the relevant product market'.[11] This changed in a fundamental way after 1980.

7.2.2 The Impact of Japanese Transplants and Lean Production Methods on the North American Automotive Labour Relations System

At the end of the 1970s, two factors conspired to initiate a fundamental shift in the North American automotive industry. First, the industry was forced to respond to new competitive conditions brought about by rising fuel costs and the increasing penetration of the domestic North American market by smaller and more fuel-efficient vehicles imported by Japanese automakers. By the mid-1980s, this competition took the form of Japanese, and later Korean and European, direct investment to construct greenfield assembly and component plants on North America soil. Second, the D-3's own competitive position had steadily eroded over the previous decade as workers continued to receive real wage increases even as labour productivity faltered. As a consequence, the D-3 were forced to restructure their operations. This included a wave of plant closings in the United States, investment in new greenfield plants such as GM Saturn, and the modernization of existing facilities through the introduction of advanced manufacturing technologies and new forms of work and production organization.

The transplants served as important 'demonstration' sites for the transfer and adaptation of Japanese management and human resource practices to a North American workforce. The transplants were built mostly in smaller rural communities and hired young workers with little or no previous industrial or union experience. Although Toyota, Honda and Nissan paid D-3 union-level wage rates to pre-empt possible unionization, they still enjoyed a considerable labour cost advantage due to the lower age-related cost of pension and health insurance benefits.

As D-3 market share declined and the economic recession of the early 1980s took hold, negotiated labour contracts in the United States reflected the weakened bargaining power of the UAW. Starting in 1979 with efforts to avert the bankruptcy of Chrysler, the traditional annual increases in real wages were replaced by lump-sum payments, profit sharing and only periodic increases in base pay. The union also agreed to concessions with regard to work rules and benefits. The move to contingent compensation schemes led to variation in wages across the D-3. Initially, these contract concessions were viewed as 'temporary' but as the pace of technological and organizational change quickened, pressure mounted to transform the traditional Fordist labour relations system into a 'new industrial relations system' (Kochan et al., 1986).

This new system was reflected in what GM referred to as 'World Class Contracts' and Ford and Chrysler as 'Modern Operating Agreements'. Such local agreements included contingent compensation in the form of lump-sum payments and profit sharing, simplified seniority provisions and reductions in job classifications, provisions for the implementation of the work

[11] Even when GM tried in the 1960s to establish a non-union, lower-wage segment of the auto industry in the southern right-to-work states, the UAW had sufficient power to gain union recognition and extend pattern bargaining to the new plants.

team concept, and increased management rights and flexibility with regard to work assignment and scheduling. As part of the transformation, the UAW agreed to modify its traditional adversarial role in favour of a new 'joint and cooperative partnership' with management. Womack et al. (1990) coined and popularized the term 'lean production' to describe a more flexible North American hybrid production system that emerged from the 1980s restructuring. Lean production grafted elements of the highly successful Toyota Production System (e.g. JIT, quality circles, kaizen and relational subcontracting) coupled with advanced computer process control equipment onto the Fordist production model (Kochan et al., 1997).

The 1980s restructuring process had a very direct bearing on labour relations in the North American industry. A significant divergence in union strategy and auto bargaining outcomes developed between the United States and Canada (Gindin, 1998; Holmes, 1991; Holmes and Rusonik, 1991; Holmes and Kumar, 1993). Faced with plant closings, layoffs and a declining membership in the United States, the UAW traded-off wage and work rule concessions and developed a more cooperative relationship with management in return for some measure of future job security.[12] In sharp contrast, UAW membership and bargaining strength was growing in Canada as a result of the stronger economic performance of the industry based on Canada's significant labour cost advantage over US plants, which has led to a situation whereby many of the D-3 models are built – often exclusively – in Canada, particularly Ontario. Canadian autoworkers firmly rejected management demands for concessions and successfully defended the traditional Fordist wage-setting formula. On ideological and political grounds, the Canadians resisted the introduction of the 'team concept' and the move to a more cooperative relationship between management and the union. This strained the relationship between the UAW's Canadian wing and the union leadership in Detroit. The rupture came in 1985 when the Canadian autoworkers left the UAW to form the autonomous CAW (Gindin, 1998; Gindin, 1995; Holmes and Rusonik, 1991). The contract settlements won by Canadian strikes in 1984 and 1987 produced fundamentally different patterns of compensation rules and work practices in Canada and the United States. The divergence that first appeared during the 1979–1983 concession bargaining period widened, so there were now two 'patterns' of labour contracts: a Canadian pattern which represented continuity with the traditional post-war Fordist pattern and a US pattern which had abandoned the latter in favour of 'modern operating agreements'. In contrast to the very strong and uniform pattern to D-3 bargaining outcomes prior to 1979, going forward, significant variations developed in wages and work practices between companies, between plants within the same company, and between the United States and Canada.

Following the 1979–1985 wave of concession bargaining in the United States, the fortunes of the D-3 improved. Even as more transplants and non-union component plants opened, the D-3 prospered during the late 1980s and 1990s, due largely to a buoyant consumer debt-fuelled market and American consumers' insatiable demand for light trucks, Sport utility vehicles (SUVs) and minivans – market segments in which the Japanese automakers had only a limited presence. By 2000, total automotive employment in the United States had risen to 905 000 from 717 600 in 1980 and the total number of vehicles produced rose in the same period from 8 009 841 to 12 773 714 (Katz, 2008: 135). In Canada, employment in assembly

[12] From 1979 to 1982, employment in the US auto industry declined 29.4% – almost 300 000 workers (Katz et al., 2002: 60). UAW contracts after 1979 began to include income and job security programmes, including 'Jobs Bank' programmes, tied to seniority. In the 1990s, these programmes were enlarged to include guaranteed employment levels at each D-3 plant (Katz et al., 2002: 73).

and parts rose from 88 427 in 1980 to 153 866 in 2000 (Statistics Canada CANSIM 281-0024) and production peaked in 1999 at 3.06 million vehicles (OICA http://www.oica.net/category/production-statistics/1999-statistics/).

However, these aggregate numbers masked a number of crucial shifts underway in the industry since the early 1980s. First, the parts sector's share of total automotive industry employment had risen significantly. This reflected increases in vehicle content, component complexity and component outsourcing as the industry moved to modular forms of assembly. Second, as Asian and European transplant production and employment expanded, the D-3 and especially GM suffered a steady loss of market share and were forced to shed employment.[13] Third, the combination of D-3 plant closings and the opening of more assembly transplants shifted the geographical locus of vehicle production away from the traditional Great Lakes auto-making region towards the south and mid-central regions of the United States.

7.3 Developments in North American Auto Labour Relations Since 2000

Since 2000, the D-3 and their unions have faced a steadily deteriorating competitive environment and an uncertain future. Growing global overcapacity, increasing competitive pressure from the transplants, and mounting financial losses, forced new rounds of D-3 restructuring and concession bargaining; Concessions that were of an order of magnitude greater than the concessions of the early 1980s. Then, in 2008–2009, the bankruptcies of GM and Chrysler triggered by the global financial crisis afforded employers as described in Chapter 11 aided in a significant and unprecedented way by the state, the opportunity to extract further deep concessions from autoworker unions in both the United States and Canada. By 2012, not only had differences in bargaining outcomes narrowed between the United States and Canada but there was a new reality in which 'union and non-union work in the [US and Canadian] auto industry have been rendered indistinguishable' (Siemiatycki, 2012).

7.3.1 Concession Bargaining 2003–2008

It is important to recognize just how much production and employment had already declined at the D-3 and their suppliers prior to the 2008 global financial crisis. In the United States, total automotive employment fell from 1 130 900 in 2000 to 827 900 in 2007(Katz, 2008: 132). The comparable numbers for Canada were 164 500 and 134 790 (Statistics Canada CANSIM 281-0024). The increase in oil prices in 2001 led to a sharp decline in sales of trucks and SUVs and a drop in D-3 market share. Union bargaining power continued to weaken as union membership dwindled at the D-3 and the UAW and CAW failed to organize the transplant assemblers or make significant inroads into the expanding non-union supplier plants (Katz, 2008: 132).

[13] In the United States, GM's market share fell from 39.2% in 1986 to 28.4% in 1999 and the number of GM hourly paid employees fell from 457 000 in 1985 to 196 000 in 2000. The loss in hourly paid employment was due in part to loss of market share but, importantly, also to significant and sustained increases in labour productivity as a result of the adoption of lean production methods.

Faced with potential bankruptcies and the threat of widespread plant closures – Ford, for example, announced in late 2005 that it would close 14 plants and significantly downsize its hourly paid labour force – the UAW agreed to reopen its 2003–2007 collective agreements with the D-3. The union agreed to benefit concessions and massive reductions in D-3 workforces largely facilitated through voluntary severance or early retirement programmes. By late 2006, 46% of Ford's hourly workforce and 31% of GM's workforce had opted to accept a buyout or retirement package (Katz, 2008: 138).

Negotiations in 2005 between the UAW and the large US automotive parts maker Delphi proved a pivotal event.[14] The concessions extracted by Delphi from the UAW became a template for 'ultra-concessionary bargaining' (Chaisson, 2012) at other suppliers and with the D-3 themselves in the period leading up to and during the 2008–2009 auto crisis (Aschoff, 2012: 137). Delphi had matched the GM contract in 1999 bargaining but, at the next round of scheduled negotiations in 2003, argued that in order to compete with non-union suppliers in the United States, it needed to significantly cut its workforce and introduce a tiered wage and benefit structure which would see new hires paid only half the wage of existing workers. The UAW agreed to the tiered wage structure in return for Delphi keeping several threatened plants open.

By the fall of 2005, Delphi was back demanding a cut in wages to $10 per hour (from the then current rate of $26 per hour), elimination of COLA and a reduction in benefits and paid vacations (Gindin, 2005). As Delphi entered a long drawn-out process of Chapter 11 bankruptcy restructuring, the UAW found itself between a rock and a hard place. An *Automotive News* editorial (Lapham, 2005) remarked that:

> Whether the unions willingly negotiate or the bankruptcy court voids Delphi's labor agreements … wages and benefits will be cut, plants will close and workers will lose their jobs. It is inevitable … once the new competitive equilibrium is set during the Delphi reorganization it will become the standard for other suppliers and the bogey for the Big 3 when they head into bargaining in 2007.

A negotiated 'Special Attrition Program' saw the sale or closure of 24 Delphi plants in the United States, 24 000 jobs cut, and allowed Delphi to make all new hires either temporary or second-tier workers. Ultra-concessionary bargaining had arrived with a vengeance in the US automotive sector.

During the 2007 round of collective bargaining with the UAW, the D-3 pushed hard to extend the Delphi-style concessions into the auto assembly sector. The UAW agreed to the establishment of a *Voluntary Employee Beneficiary Association* (VEBA), which effectively removed the cost obligation for retiree medical benefits from the automakers' books.[15] The union also accepted second-tier wages and health and pension benefits for new 'non-core' hires – those in sub-assembly, machining and material handling. The $14–16 per hour starting wage rates for these workers were approximately 50% of previous rates and new hires were

---

[14] GM had spun-off Delphi, its in-house parts manufacturing division, into an independent corporate entity in 1998. The UAW had continued to represent Delphi workers who at the time of the spin-off were covered under the GM Master Agreement.

[15] The VEBA is known as the UAW Retiree Medical Benefits Trust. The Trust is controlled by a board of UAW appointed and independent directors and provides healthcare benefits for GM, Chrysler and Ford retirees. GM alone had an estimated $51 billion in retiree healthcare obligations on its books in 2007 which it shifted to the VEBA.

excluded from the defined benefit pension plan and the VEBA. It was reported that these concessions would cut the overall cost of a GM new hire to about one-third that of existing workers (Ward's Automotive, 22 October 2007).

The 2007 round of bargaining narrowed the gap in all-in labour costs between D-3 plants and the transplant automakers.[16] It also reduced the labour cost advantage enjoyed by D-3 plants in Canada and set the stage for the D-3 to apply pressure on the CAW to grant concessions in scheduled bargaining the following year. In April 2008, and perhaps cognizant of the impending financial crisis, the CAW negotiated a new contract with Ford Canada a full 5 months ahead of the expiry of the 2005–2008 agreement. The new agreement included a number of significant concessions: one-time 'bonus' payments in lieu of a freeze on wages for the duration of the 3-year agreement, a 2-year freeze on COLA, a cut in paid vacation time and reduced pension medical entitlements.[17] It also contained a commitment by the parties to establish a VEBA-type pre-funded health benefit fund for retirees. While the CAW leadership made much of the fact that they had successfully resisted a UAW-style two-tier wage structure for new hires, they did agree to a 'New Hire Grow-in System' under which new hires would start at 70% of base wages and benefits and only attain 100% after 3 years. The deal with Ford was quickly replicated at GM and Chrysler.

7.3.2 The Impact of the Global Financial Crisis

While 2007 and 2008 were difficult and challenging bargaining rounds for the UAW and CAW, much more was yet to come. In late 2008, as the global financial crisis deepened and credit markets froze, automobile sales in the United States plummeted to a 40 year low. The D-3 automakers entered the crisis in an already fragile situation, by virtue of several years of shrinking domestic sales. Stanford (2010) observed that:

> .... anti-union public attitudes, hostile labour laws and the failure of unions to gain a toehold at the transplant facilities further contributed to the political isolation of the North American OEMs and their unions. When government support eventually came, it was delivered in a manner that took advantage of the schism between the North American owned unionized plants and the offshore-owned non-union plants.

To continue operating, GM and Chrysler were forced to seek financial assistance from the US government. Opposition to Bush's GM and Chrysler rescue package was led by US Senate representatives from southern right-to-work states who 'did not want the US federal government to directly assist the American-owned companies that were the main rivals of the Japanese OEMs that had invested heavily in their own states; regional loyalty, in this sense, overwhelmed whatever "national" loyalty might have still existed towards the American-

[16] While the VEBA had a significant impact on the cost competitiveness of the D-3 versus the transplants (and versus D-3 plants in Canada), initially, the new hire rate had less impact since the D-3 were not in hiring mode.

[17] The appreciating value of the Canadian dollar after 2002 steadily eroded Canada's cost advantage vis a vis US auto plants and the CAW began to make localized concessions on work rules and outsourcing in order for Canadian plants to compete against plants in the United States for new product mandates (Siemiatycki, 2012). But it was not until 2008 that the CAW agreed to contingent compensation.

owned OEMs' (Stanford, 2010: 398). After a lengthy process of congressional hearings, the Bush administration granted initial federal loans to the companies on 19 December 2008.

Additional federal assistance was made conditional on each company providing the US Treasury Department with a restructuring plan that would ensure the long-term viability of the company. In late March 2009, the newly installed Obama administration rejected GM's initial restructuring plan, which included an agreement from the UAW to reopen the 2007 collective agreement and make further concessions, and required the automakers to make 'every effort to achieve labor cost parity with the [non-union] transplants' (USGAO, 2009: 26). The UAW National Negotiating Committee's message to GM workers in May 2009 urging ratification of the concessionary agreement reflects the extraordinary degree to which the state intervened into collective bargaining:

> [after Obama's rejection of GM's restructuring plan] The Treasury's Auto Task Force moved into GM headquarters in an effort to find deeper and quicker changes to GM's US footprint. The shocking plan that resulted, highlighted by the elimination of 16 existing US facilities ... levied severe and drastic cuts to the US workforce and UAW membership. Faced with this dire situation and realizing that failure to meet the government requirements would surely mean the end of General Motors, your bargainers painstakingly put together modifications to the [2007] collective bargaining agreement to satisfy the Treasury Auto Task Force.

The bailout concessions greatly expanded those made in 2007 by extending second-tier wages to all new hires, reducing paid-time off and skilled trades wage premiums, suspending the job security programme, cutting SUB benefits and loosening restrictions on the use of temporary workers. Wages for entry-level workers were frozen, COLA was suspended and GM and Chrysler workers were barred from striking for 6 years. The VEBA, negotiated in 2007, was modified to reduce retiree benefits.

Although GM Canada and Chrysler Canada did not formally file for restructuring under bankruptcy protection in Canada and the overall auto bailouts were orchestrated by the US Treasury Department, the Canadian federal and Ontario provincial governments were major lenders to GM and Chrysler and also took equity holdings in both companies. It is highly probable that the US government would have bailed out GM and Chrysler with or without assistance from Canada. For Canada, however, the issue was not whether GM and Chrysler would survive but rather whether they would continue to have operations in Canada.[18] In the absence of Canadian government involvement, it is likely that GM and Chrysler would have been under great political pressure to close their Canadian operations and retrench production in the United States.

Canadian governments followed the Obama administration's lead and made financial aid conditional on labour concessions that went beyond those already negotiated with the companies (Stanford, 2010). The final concessions included further reductions in paid-time off, limits on the SUB programme, reduced supplementary medical benefits and changes in local plant operating rules to increase productivity. In contrast to the UAW concessions, however, the CAW retained the right to strike in their next round of scheduled contract negotiations (Stanford, 2012: 399).

[18] Total US and Canadian government assistance amounted to approximately $100 billion of which about $80 billion was directed to GM. Governments in Canada provided roughly 20% of the total funding and, in return, received commitments from each company to maintain their then current share of total North American production in Canada (Stanford, 2012: 399).

After achieving new labour agreements with the UAW and CAW and with political and financial support from the US and Canadian governments, GM and Chrysler entered an accelerated bankruptcy restructuring and were reborn a few weeks later as 'new' smaller but leaner and financially stronger companies. Although Ford had not required bankruptcy restructuring or government aid, both the UAW and CAW extended the GM and Chrysler contract concessions to Ford, and in Canada, to CAMI.

7.3.3 *Post-Crisis Developments*

Even though auto sales had picked up, the UAW, conscious of divided public sentiment over the taxpayer funded bailouts and with no right to strike at either Chrysler or GM, entered the 2011 scheduled D-3 contract negotiations with very modest goals. The union achieved small improvements in benefits and base wages rates for entry-level workers, but other base wage rates and COLA remained frozen. In seeking membership ratification of the agreement, the UAW leadership stressed their success in securing commitments from each company for new investment at a number of plants and the repatriation to the United States of some off-shore auto assembly work.

In the lead-up to the 2012 round of bargaining in Canada, the D-3 launched a concerted media campaign emphasizing that Canadian labour costs were now significantly higher than in the United States.[19] The CAW admitted that labour costs were now 'modestly higher' in Canada due both to the continued appreciation of the Canadian dollar and the impact of the UAW second-tier wage and benefit concessions. The D-3 publically threatened disinvestment from their Canadian plants if the cost gap with the United States was not closed. The union 'signalled it would contemplate forms of compensation improvements (such as lump sum benefits) that would not increase the fixed cost gap between Canadian and US plants. It also indicated a willingness to negotiate around the terms of the new-hire grow-in programme [first negotiated in 2008]' (Stanford, 2012). The new 4-year agreement included a freeze on base wages and suspension of COLA, $9000 in lump-sum bonuses, a freeze on pensions and incremental improvements in some benefits.

The aspect of the CAW agreements that attracted most critical comment was the provision that new hires would now begin at 60% of base wages and only attain 100% after 10 years and would receive a new hybrid pension plan instead of the traditional defined benefit plan. So, in essence, while the companies' demand for US-style 'permanent' two-tier wages and benefits for new hires was resisted, the CAW did accept 'temporary' 10-year tiered wages and benefits (Fowler, 2013).

7.4 Conclusion

Auto industry labour relations in the United States and Canada are still conducted within essentially the same legal framework that emerged from the institutional restructuring of the 1930s. The labour relations landscape, however, has been significantly reshaped over the course of the last 30 years as the North American auto industry underwent a massive transition.

[19] Ford claimed a total cost gap of $15 per hour and Chrysler argued that costs at its Canadian plants were 20% greater than in the United States.

Prior to 1980, the assembly sector was fully unionized and there was a high level of union coverage in the components sector. Workers across the industry in both the United States and Canada were represented by one large and powerful industrial union, the UAW. With only limited import penetration by foreign automakers, the geographical reach of union organization closely matched the extent of the production system. There was a well-functioning labour relations system that underpinned long-term growth in production and employment and produced rising real income for autoworkers whose earnings were significantly higher than the average manufacturing wage.

During the 1980s, first Japanese, and later European and Korean, companies began producing vehicles in North America, dramatically increasing competition in the industry and challenging the Big Three. However, 'over the 1980s and 1990s, even when the fortunes of the UAW and the Big Three turned down, [their] influence on the US economy and American collective bargaining remained substantial' (Katz, 2008: 132).

After 2000, the D-3 and their unions (UAW and CAW) faced a growing crisis precipitated by their continued loss of market share to transnational automakers and through their failure to organize the expanding transplant assembly sector or make significant inroads into the non-union component sector.[20] The 2000s saw a substantial decline in employment in D-3 assembly plants and their key suppliers, significantly reduced wages and benefits for new hires, and numerous other contract concessions. By 2008, there were 435 000 fewer jobs in the US motor vehicle manufacturing sector as compared to 2000.

Although the recession triggered by the 2008 global financial crisis hit the auto industry hard in all regions of the world, production and employment declined more sharply in North America than in virtually any other jurisdiction. The D-3 closed 17 car and light truck assembly plants between May 2007 and 2011 in the United States, and GM's hourly paid labour force was cut from 61 000 in 2008 to 40 000 in 2010 (Platzer and Harrison, 2009).

The historic state-mandated concessions made by the UAW and CAW during the 2009 bailout crisis, together with the concessions made in 2007 and 2008, significantly narrowed the gap in all-in labour costs between the D-3 and the long established Toyota and Honda plants in Ohio, Kentucky and Ontario (see Stanford, 2012). Newer transplants in southern US right-to-work states, such as the recently opened VW and KIA assembly plants in Tennessee and Georgia, have wage rates in the $14–20 per hour range. The restructuring significantly reshaped the geography of the North American auto industry. Today, there are essentially two distinct regional clusters: a northern Great Lakes region composed of what remains of unionized D-3 production and a southern US region characterized by lower wage non-union transplant producers (Klier and Rubenstein, 2010).

So, what are the future prospects for auto industry industrial relations in North America? After sliding precipitously and hitting a low point in 2009 employment in the industry has stabilized and as North American vehicle sales steadily recover, by 2014, assembly plants that survived the latest restructuring were running at close to full production capacity. The 'new' slimmed-down D-3 announced new investments to expand capacity, largely in the US Great

[20] Most recently (February 2014), workers at the VW plant opened in 2011 in Chattanooga, Tennessee, voted 53:47% to reject UAW representation (Trotman and Maher, 2014). This represents a major setback for the UAW because the company had remained genuinely neutral during the unionization drive. Following the failed union certification vote, VW indicated that it would like to see a German-style works council established in the plant – this would represent a major innovation in North American automotive industrial relations but is difficult to institute under current US labour law.

Lakes region (Center for Automotive Research, 2013). However, union membership in the auto sector has been reduced to a shadow of its former self and, as more transplants come on stream in the US South and Mexico, unionized assembly plants now account for barely half of total vehicle output in North America.[21] The significant wage premium over the manufacturing average that autoworkers used to enjoy has been greatly reduced (Andes and Muro, 2013). Michigan and Indiana – states in the historic heartland of the US auto industry – recently enacted right-to-work legislation undermining the ability of unions to organize and represent workers. Given these trends, it is unlikely that the UAW and CAW will ever regain either the membership or the considerable power and influence they once enjoyed in shaping auto industry labour relations and employment practices.

References

Aglietta, Michel, (1979) *A Theory of Capitalist Regulation: The US Experience*. London: New Left books.

Anastakis, Dimitry, (2005) *Auto Pact: Creating a Borderless North American Auto Industry, 1960–1971*. Toronto: University of Toronto Press.

Anastakis, Dimitry, (2013) *Autonomous state: The Struggle for a Canadian Car Industry from OPEC to Free Trade*. Toronto: University of Toronto Press.

Andes, Scott and Mark Muro, (2013) *Next for the U.S. Auto Sector: Beyond Wage Competition*, Washington, DC: Brookings Institute, The Avenue October 9.

Aschoff, Nicole, (2012) A tale of two crises: labour, capital and restructuring in the US auto industry, *Socialist Register*, **48**, 125–148.

Boyer, Robert, (1990) *The Regulation School: A Critical Introduction*. New York: Columbia University Press.

Center for Automotive Research, (2013) *The Automotive Corridor: How Automotive Investment in the Great Lakes Is Shaping the Assets, Institutions, and Infrastructure of the Region, Creating a World Class Advanced Manufacturing Environment*. Ann Arbor: Center for Automotive Research.

Chaison, Gary, (2012) *The New Collective Bargaining*. New York: Springer.

Dicken, Peter, (2011) *Global Shift*. New York: Guilford.

Edwards, Tom, Tom Colling and Anthony Ferner, (2007) Conceptual approaches to the transfer of employment practices in multinational companies: an integrated approach, *Human Resource Management Journal*, **17**(3), 201–17.

Fowler, Tim, (2013) Does fighting back still matter? The Canadian autoworkers, capitalist crisis and confrontation, *Capital & Class*, **36**(3), 493–513.

Gindin, Sam, (1995) *The Canadian Autoworkers Union: The Birth and Transformation of a Union*. Toronto: James Lorimer.

Gindin, Sam, (1998) Breaking away: the formation of the Canadian Autoworkers Union, *Studies in Political Economy*, **29**, 63–89.

Gindin, Sam, (2005) GM, the Delphi Concessions, and North American Workers: Round Two? *MRzine*, 15 November. New York: Monthly Review. Available at http://mrzine.monthlyreview.org/2005/gindin151105.html. Accessed 26 March 2015.

Holmes, John, (1991) From uniformity to diversity: changing patterns of wages and work practices in the North American Automobile industry. In P. Blyton and J. Morris (eds), *A Flexible Future? Prospects for Employment and Work Organization in the 1990s*. p. 129–151. Berlin: De Gruyter.

Holmes, John, (1993) From three Industries to one: towards an integrated North American Automobile Industry. In M. Appel Molot (ed), *Driving Continentally: National Policies and the North American Auto Industry*. p. 23–62. Ottawa: Carleton University Press.

[21] Of the 10 new assembly plants opened or announced in the US between 1997 and 2012, 8 were in right-to-work states and the other 2 were in Indiana, which enacted right-to-work legislation in 2012 (Klier and Rubenstein, 2013: 12). Auto industry growth in Mexico has been phenomenal since 2008 and Mexico now produces more vehicles than Canada. Mexico has signed numerous free trade agreements and with extremely low labour costs and high quality is increasingly an export platform for Latin American countries as well as North America.

Holmes John and Pradeep Kumar, (1993) Labour Movement Strategies in the Era of Free Trade: "Competitiveness" and the Uneven Transformation of Labour-Management Relations in the Canadian and U.S. Auto Industries. In Rianne Mahon, Jane Jenson and Manfred Bienefeld (eds), *Production, Space, Identity: Canadian Political Economy Faces the Twenty-First Century*. p. 195–224. Toronto: Canadian Scholars Press.

Holmes John and Anthony Rusonik, (1991) The break-up of an international labour union: uneven development in the North American auto industry and the schism in the UAW. *Environment and Planning A*, **23**(1), 9–35.

Holmes John and Tod Rutherford, (2013) After the Great Recession: Labour Regulation and Collective Bargaining in the United States and Canada. Paper presented at the Annual Meetings of the American Association of Geographers, Los Angeles, USA. April 2013.

Hunter, Larry and Harry Katz, (2012) The impact of globalization on human resource management and employment relations in the US automobile and banking industries, *International Journal of Human Resource Management*, **23**(10), 1983–1998.

Katz, Harry, (1985) *Shifting Gears: Changing Labor Relations in the US Automobile Industry*. Cambridge: MIT Press.

Katz, Harry, (2008) Recent developments in US auto labor relations: the decline of the Big Three and the United Automotive Workers. In R. Blanpain and R. Lansbury (eds), *Globalization and Employment Relations in the Auto Assembly Industry*. Bulletin of Comparative Labour Relations **64**. p. 131–142. The Hague: Kluwer Law International.

Katz, Harry and Owen Darbishire, (2000) *Converging Divergencies: World Wide Changes in Employment Systems*. Ithaca: Cornell University Press.

Katz, Harry, John Paul MacDuffie and Fritz Pil, (2002) Autos: continuity and change in collective bargaining. In P. Clark, J. Delaney and A. Frost (eds), *Collective Bargaining in the Private Sector*. . p. 55–90. Champaign: Industrial Relations Research Association

Kirsch, Anja and Nick Wailes, (2012) Varieties of employment relations: continuity and change in the global auto and banking industries, *International Journal of Human Resource Management*, **23**(10), 1967–1982.

Klier, Thomas and James Rubenstein, (2010) The changing geography of North American motor vehicle production, *Cambridge Journal of Regions, Economy and Society*, **3**, 335–347.

Klier, Thomas and James Rubenstein, (2013) Restructuring of the U.S. auto industry in the 2008–2009 recession, *Economic Development Quarterly*, **7**(2), 144–159.

Kochan, Thomas, Harry Katz and Robert McKersie, (1986) *The Transformation of American Industrial Relations*. New York: Basic Books.

Kochan, Thomas, John Paul MacDuffie and Russell D. Lansbury (eds), (1997) *Responding to Lean Production: Emerging Employment Practices in the World Auto Industry*. Ithaca: Cornell University Press.

Kumar, P. and J. Holmes, (1996) Change but in what direction?: divergent union responses to work restructuring in the North American auto industry. In F. Deyo (ed), *Social Reconstructions of the World Automobile Industry: Competition, Power and Industrial Flexibility*. p. 159–199. London: International Political Economy Series, MacMillan.

Kumar, Pradeep and John Holmes, (1997) Canada: continuity and change. In T.A. Kochan, J.P. MacDuffie and R.D. Lansbury (eds), *Responding to Lean Production: Emerging Employment Practices in the World Auto Industry*. p. 85–108. Ithaca: Cornell University Press.

Lansbury, Russell, Jim Kitay and Nick Wailes, (2003) The impact of globalization on employment relations: some research propositions, *Asia-Pacific Journal of Human Resources*, **41**(1), 62–74.

Lansbury, Russell, Nick Wailes, and Anja Kirsch, (2008) Globalization, continuity and change: the automotive assembly industry. In R. Blanpain (ed), *Globalization and Employment Relations in the Auto Assembly Industry*. Bulletin of Comparative Labour Relations **64**. p. 143–152. New York: Wolters Kluwer.

Lapham, Edward (2005), Delphi's wage war will reshape U.S. labour market, *Automotive News*, 8 November 2005.

Lipietz, Alain, (1987) The globalization of the general crisis of Fordism, 1967–84. In J. Holmes and C. Leys (eds), *Frontyard/Backyard: The Americas in the Global Crisis*. p. 95–129. Toronto: Between the Lines Press.

Platzer, Michael and Glennon Harrison, (2009) *The US Automotive Industry: National and State Trends in Manufacturing Employment*. Washington, DC: Congressional Research Service. Available at http://digitalcommons.ilr.cornell.edu/key_workplace/666. Accessed 26 March 2015.

Rutherford, Tod and John Holmes (2013) (Small) differences that (still) matter? Cross border regions and work place governance in the Southern Ontario and U.S. Great Lakes automotive industry, *Regional Studies*, **47**(1), 116–127.

Siemiatycki Elliot (2012), Forced to concede: permanent restructuring and labour's place in the North American auto industry, *Antipode*, **44**(2), 453–73.

Smith, Chris, Brendan McSweeney and Robert Fitzgerald (eds), (2008) *Remaking Management: Between Global and Local*. Cambridge: Cambridge University Press.

Sorge, Arndt (2005) *The Global and the Local: Understanding the Dialectics of Business Systems*. Oxford: Oxford University Press.

Stanford, James, (2010) The geography of auto globalization and the politics of auto bailouts, *Cambridge Journal of Regions, Economy and Society*, **3**, 383–405.

Stanford, James, (2012) Canadian auto workers deliver acceptable deal after threat to strike Detroit 3, *Labor Notes*, 16 October 16 (consulted 25 May 2013). Available at http://www.labornotes.org/blogs/2012/10/canadian-auto-workers-deliver-acceptable-deal-after-threat-strike-detroit-3?language=en. Accessed 26 March 2015.

Sturgeon, Timothy, Johannes Van Biesebroeck and Gary Gereffi, (2009) Value chains, networks and clusters: reframing the global automotive industry, *Journal of Economic Geography*, **8**, 297–321.

Trotman, Melanie and Kris Maher, (2014) Volkswagen vote loss signals difficulty for union organizers, *Wall Street Journal*, 15 February 2014. Available at: http://www.wsj.com/articles/SB10001424052702304315004579385372001325260. Accessed April 18, 2015.

United States Government Accountability Office (USGAO), (2009) Summary of Government Efforts and Automakers' Restructuring to Date, Report to Congressional Committees, GA-09-553, Washington, DC.

Wailes, Nick, Russell Lansbury, Jim Kitay and Anja Kirsch, (2008) Globalization, varieties of capitalism and employment relations in the automotive assembly industry. In R. Blanpain (ed), *Globalization and Employment Relations in the Auto Assembly Industry*. Bulletin of Comparative Labour Relations **64**. New York: Wolters Kluwer.

Womack, James, Dan Jones and Dan Roos (1990) *The Machine That Changed the World*. New York: Rawson Associates.

8

Labour Relations and HRM in the Automotive Industry: Japanese Impacts

Katsuki Aoki
School of Business Administration, Meiji University, Tokyo, Japan

8.1 Introduction: The Japanese Car Industry and Toyota Production System

In 1990, when the book *The Machine That Changed the World* (Womack et al., 1990) was published, Japan was the largest vehicle (car and truck) producer in the world, whose production volume was 13 486 796, superior to the United States, the second largest producer. In 2012, Japan produced 9 942 793 cars and was only the third largest producer, fewer than China and the US's production numbers of 19 271 808 and 10 328 884, respectively. Although car production in Japan decreased by about 3.5 million between 1990 and 2012, this does not mean that the global presence of Japanese car makers decreased. As Figure 8.1 shows, Japanese car makers significantly increased their overseas production in the same period. In 2012, overseas production volume by Japanese makers amounted to 15 825 398, about 150% of their domestic production volume. When we look at each maker's production volume in 2012, as Figure 8.2 shows, Toyota stands out among Japanese car makers, whose production volume share was almost one-third of total global production by Japanese makers.

Toyota employs its world-leading production system, known as the Toyota Production System (TPS). Mainly in the 1980s, this system began to attract much attention from the wider world. In a narrow sense, this is a system that calls for the right parts to be supplied only when they are needed and only in the amount needed. However, in a broader sense, this system includes quality circles, suggestion schemes and teamwork that facilitate the organization to improve itself; this

The Global Automotive Industry, First Edition. Edited by Paul Nieuwenhuis and Peter Wells.

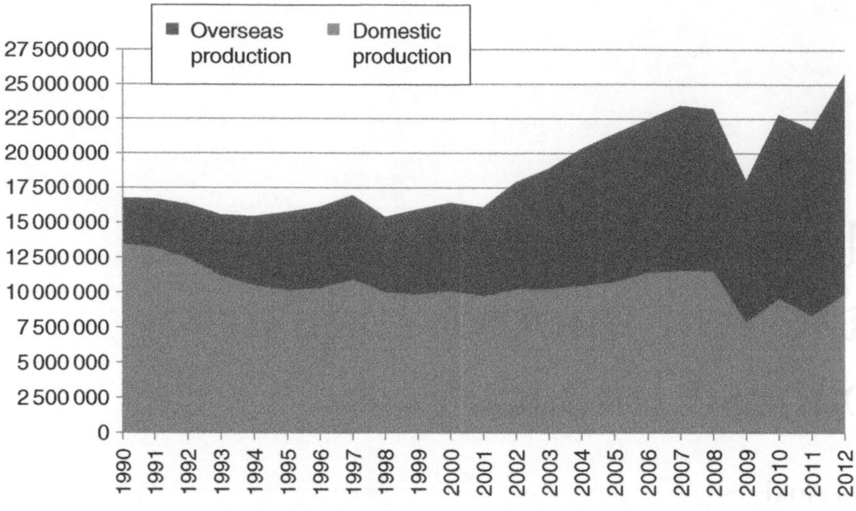

Figure 8.1 Domestic and overseas car production by Japanese makers. Source: Compiled from data supplied by Japan Automobile Manufacturing Association, Inc. *Nihon no Jidosha Kogyo [Japanese Automobile Manufacturing]*

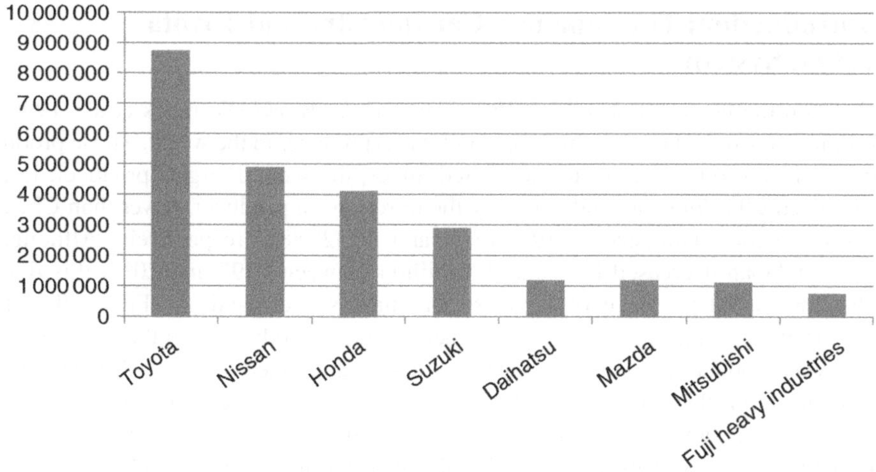

Figure 8.2 Global production volumes by major Japanese car makers in 2012. Source: Compiled from individual company's official webpages

is called 'kaizen' or continuous improvement. These practices have been developed based on unique Japanese human resource management (HRM) practices and labour relations, such as lifetime employment, seniority-based pay and collaborative industrial relations.

The purpose of this chapter is to follow the development of TPS in a broader sense and Japanese labour relations and HRM as enablers for implementing TPS, thereby examining the Japanese impacts on the global automotive industry. Firstly, in Section 8.2, TPS is briefly explained, after which the roles of Japanese labour relations and HRM in the implementation of TPS are looked at. Then, in Section 8.3, the impacts of TPS on the global automotive industry are described under the name of 'Japanization'. Finally, in Section 8.4, changes in Japanese labour relations and HRM after the bubble economy burst and their impacts on the implementation of TPS are focused on. In the conclusion section, problems and future challenges for TPS and the Japanese automotive industry are described in relation to current changes in the global automotive industry.

8.2 TPS and Japanese HRM

TPS consists of the following two concepts (Ohno, 1988): One is 'autonomation' (automation with a human touch) that was developed by Sakichi Toyoda around 1900. This concept means that machines automatically detect problems and stop if problems occur, thereby preventing the production of defective products: *Pokayoke*, which shares this basic concept, is well known in the case of automotive production. The other one is Just-in-Time (JIT) which was conceived by Kiichiro Toyoda and applied to Toyota Motor Co.'s Koromo Plant starting production in 1938. This concept means that manufacturers produce only what is needed, when it is needed and in the amount needed. These two concepts evolved into TPS owing mainly to the efforts of Taiichi Ono after World War II.

TPS represents a further development of the Ford production system (or Ford–Budd system, as outlined in Chapter 4) that facilitated the transition from mass production to small lot sizes, reducing the cost of holding large inventories of the parts required on conveyor-based assembly lines (see also Chapter 5). In this production system, the right parts are supplied to the production line only at the time they are needed and only in the amount needed. This system relies on *kanban* that is used for material flow control to ensure effective communication between internal and external operations on issues such as adjusting inventory levels, production schedules and lead times for the delivery of parts (Monden, 1998). *Heijunka* (production levelling) also plays a key role in this system with the aim of levelling out the quantity of parts that are sent up the supply chain, thereby smoothing flow on the production line (Monden, 1998).

From the technical point of view, *kanban* and *heijunka*, together with *pokayoke*, are important tools for implementing TPS. However, the purpose of implementing TPS is not only to reduce stocks and costs. We need to note softer issues, such as developing shared values among employees and influencing behaviour patterns, to understand TPS more deeply. *Kaizen* or continuous improvement (Imai, 1986, 1997), denoting always trying to find problems, pursue the root cause of the problems, and take and standardize the countermeasures, is one of the important purposes for implementing TPS. *Genchi-genbutsu*, roughly 'actual workplaces' and 'actual things' in English, is also considered to be an important concept that underlies people's behaviour in implementing TPS (Liker, 2004).

The actual implementation of these concepts can be observed in Japanese shopfloor work practices, typical ones include quality control (QC) circles, suggestion schemes and 5S (Cole, 1994; Imai, 1986; Schonberger, 1982). These practices can briefly be explained as follows:

1. QC circles: Regular-based voluntary problem-solving activities by shopfloor workers, the theme of which is decided by participants themselves. Not only countermeasures to solve problems but also collaborative attitudes and teamwork are aimed to be created. Circle activities are implemented based on the plan-do-check-act (PDCA) cycle using QC stories.

2. Suggestion schemes: Suggestions are provided by shopfloor workers (sometimes including indirect staff) that are used for both creative and preventive purposes. In usual cases, a certain number of suggestions, such as one idea per month, is given to employees as a target. Monetary rewards are given to employees who provide an effective suggestion.
3. 5S consists of five Japanese words starting with S: *Seiri* (clear out), *Seiton* (configure), *Seiso* (clean and check), *Seiketsu* (conformity) and *Sitsuke* (custom and practice) (as translated by SMMT Industry Forum in the United Kingdom). Usually, this is implemented by shopfloor workers as part of everyday operations. The progress is regularly checked.

In fact, these practices are not rooted in Toyota. For example, QC circles were initiated by Kaoru Ishikawa as part of promoting Japanese company-wide QC in which all members of the organization participate (Ishikawa, 1985). However, these practices are considered to be indispensable factors for implementing TPS. If we try to implement JIT and minimize buffer stocks between different processes, the quality system should be built-in on the shopfloor. These three practices are needed for creating the built-in quality system by enhancing shopfloor operators' quality consciousness and developing workplaces in which problems are easily noticeable. It can be said that seamless lean production flow created by using *kanban* depends on these three practices that support the built-in quality system.

These three Japanese shopfloor practices are closely connected to Japanese labour relations and HRM (Kenney and Florida, 1988; Womack et al., 1990). As Cole (1994) and Koike (1994) note, On the Job Training (OJT) plays a key role in implementing Japanese shopfloor practices. In the cases of QC circles and suggestion schemes, shopfloor workers need to be educated not only about analysing quality problems, but also about improving their production processes. In the case of 5S, they also need to understand the importance of always keeping the shopfloor and equipment clean and organized. In Japanese organizations, such knowledge is developed mainly through doing day-to-day operations and communication between colleagues, or OJT in short, and takes a long time to be achieved. A job rotation system, in which workers are asked to perform different jobs, also gives an opportunity for workers to better understand the relationship between their job and other jobs.

OJT and job-rotation are based on Japanese HRM, or long-term employment. As already mentioned, it takes a long time to implement OJT because it depends on daily communication and trustful relationships between colleagues. Such relationships can only develop gradually between people who work for the same organization for a long time. If employees could gain some benefits by jointly contributing to the same organization with other members, they would be willing to share important knowledge and skills with other members. However, if employees easily change companies or try to use their knowledge only for their own benefit, they would not be willing to share knowledge and skills with other members. Therefore, in Japanese companies, human development systems are linked to long-term employment and internal promotion systems that encourage collaborative action within the organization (Dore, 1973).

As Table 8.1 shows, Toyota provides its shopfloor workers with opportunities to upgrade their skills and qualifications within the company and allows them a long time to achieve this. This job grade system is linked to job positions; EX in Table 8.1 corresponds to team leader, SX to group leader and CX to chief leader. Basically, it takes at least 10 years for new workers to become group leaders. It is possible for Toyota's shopfloor workers to be further promoted to *kacho* (manager), *jicho* (deputy director) and *bucho* (director). By providing opportunities

Table 8.1 Toyota's job grade system for skilled workers

Job grade	Examples of expected skills	Wage grade
CX	Those who are able to take on a new challenge by understanding the job and environmental situation, and produce a positive result by getting related people involved	Grade 1
SX	Those who are able to develop the capability of their own group, and also to participate in a project and achieve positive results	Grade 2
EX	Those who are able to manage problems, to develop preventive measures, and to implement further improvement	Grade 3 Grade 4
Mid-level skilled workers	Those who are able to find short-term solutions to problems, and also to improve their work processes	Grade 5 Grade 6
Primary skilled workers	Those who are able to identify problems and predict dangers, and also to report these things to their boss	Grade 7 Grade 8
Basic skilled workers	Those who are able to observe the rules, and to carry out their own roles at work	Grade 9

Note: Compiled from data supplied by Saruta (2007) and Sugiyama (2004).

for promotion inside the company, Toyota's employees can be motivated not only to better understand their jobs but also to develop their skills inside the company. This grade system is also linked to wage grades, as shown in Table 8.1, and payment systems that also work for motivating workers.

Table 8.2 shows Toyota's payment system for skilled workers and average pay amount in 2003. Here, we can recognize that employees' payment in Toyota in 2003 was decided by considering different payment items. There were different methods for wage determination in the different payment items. For example, in the item of basic-ability pay for skilled workers, grade 1 people in Table 8.1 were paid almost 3 times more than grade 9 people (Sugiyama, 2004). In the case of age-based pay, people aged 52–55 were paid about 2.8 times more than people aged 18 (Sugiyama, 2004). This shows Toyota's payment was partly decided on the basis of skill grade, and partly on the basis of seniority. However, this payment system also included wage discrimination based on a personnel evaluation. As for individual ability-based pay, there was a percentage range of 85–115 for the payment of skilled workers at any grade (Sugiyama, 2004).

In Toyota, as well as other large Japanese companies, there is a negotiation between management and workers regarding employees' payment as well as other conditions, which is called the annual spring labour offensive. Here, it is rare that major conflicts between management and workers happen because large Japanese companies usually have their own company union. In the case of company unions, both management and workers share the same fate within the same company. Furthermore, usually only regular employees are members of the union, and non-regular, casual employees who are considered to be more frustrated with their conditions are not members. In general, industrial relations in large Japanese companies are collaborative, which also helps the employees develop their knowledge and skills bringing more benefits to the organization.

Table 8.2 Toyota's payment system for skilled workers in 2003

Payment items		Monthly JPY	Comments
Regular payment	Basic ability-based pay	93 013	Payment amount is decided by wage grade
	Individual ability-based pay	137 444	Payment amount is decided by individual ability
	Productivity-based pay	54 071	Basic grade-based payment × plant performance rate
	Age-based pay	56 104	Payment amount is decided by age
	Other benefits	13 397	
Irregular payment	Overtime pay	65 257	
	Night shift pay, etc.	28 334	
		Half year JPY	
Bonus	Basic bonus	768 000	Monthly regular payment × 2.207 in Summer 2003
	Grade-based amount	512 000	Grade-based amount
	Additional amount		Payment amount is decided by individual performance

Note: Compiled from data supplied by Gankoji (2005) and Sugiyama (2004).

Abegglen (1958), who based his field studies on Japanese factories, regards lifetime employment, seniority payment and company unions as the main characteristics of Japanese labour relations and HRM; these are known as *sanshu-no-jingi* (three sacred treasures). During a period of strong economic growth in Japan, a lot of research had shown that the three sacred treasures directly or indirectly contributed to Japanese companies' success, through contributing to high labour productivity (Abernathy et al., 1981), the implementation of TQM and quality circles (Cole, 1994), and continuous improvement or *kaizen* (Womack et al., 1990). However, these characteristics began to face strong pressures to adapt to more marked-oriented styles, as in the United States. Before mentioning such changes in Section 6.4, the impact of TPS and Japanese HRM on the global automotive industry will be examined in the next section.

8.3 'Japanization' of the Global Automotive Industry

According to Oliver and Wilkinson (1992), the term 'Japanization' (also 'Japonization' in some sources) generally includes the following two meanings: one is the emulation of Japanese industrial practices by Western companies and the other is the increase in Japanese FDI (Foreign Direct Investment) into Western economies (Oliver and Wilkinson, 1992). Concerning the latter, JETRO (Japan External Trade Organization) shows that Japanese FDI to the European Union was US$ 1 534mn in 1985 and increased rapidly to $11 027mn in 1990. After the Japanese bubble economy burst, it declined to less than $3 000mn in 1994, 1997 and 1998, but bounced back to $10 968mn in 2000, and doubled to $22 939mn in 2008 (JETRO, 2014).

In the discussion related to Japanization, researchers have examined not only how TPS and Japanese shopfloor work practices are implemented on Western shopfloors (see Chapter 7 and also cases in the United States by Kenney and Florida, 1993, and cases in Europe by Kumon and Abo, 2004), but also social and political issues related to the implementation of such activities overseas, such as the resistance from labour unions (see Chapter 7 and also the case of Mazda in the United States by Fucini and Fucini, 1990) and the misunderstanding between Japanese and local people (see the cases of Japanese transplants in the United Kingdom by Delbridge, 1998 and Elger and Smith, 2005). Some cases of successful transfer of Japanese shopfloor *kaizen* activities have been illustrated mainly in the United States (Kenney and Florida, 1993). The case of NUMMI, a joint plant between GM and Toyota, is considered to be one such example (Adler, 1999).

NUMMI began operation in California in 1984 (and closed in 2009), 85% of whose workers remained from the GM-Fremont plant that closed in 1982. According to Adler (1999), the productivity and quality of the Fremont plant were among the worst in the whole GM group, whose absenteeism rate was about 25% in 1982. The new NUMMI plant established the NUMMI Production System that brought in a lot of concepts and tools from TPS, such as JIT using *kanban, heijunka, kaizen*, visual control, teamwork and standardized work. This plant improved not only its plant performance but also the work conditions. Its uncorrected productivity (h/unit) in 1986 improved to 20.8 compared to 43.1 in the GM-Fremont plant in 1978; product quality (Consumer Report Reliability Index) also improved in the same period from 2.6–3.0 to 3.6–3.8 (Adler, 1999). Regarding work conditions, the absenteeism rate was down to 2.5% from 25%; the number of grievances was down to only around 30 compared to over 700 in 1982.

NUMMI also introduced some elements from Japanese HRM (Kenney and Florida, 1993). For example, production teams in this plant had a daily meeting to discuss production improvements and task redesigns. This plant also rotated workers in the same team frequently. Kenney and Florida's (1993) research shows that the US big three neither had such work teams nor rotation. The situation of NUMMI in terms of work teams and rotation was similar to that of other Japanese transplants in the United States, such as Toyota, Honda and Nissan. Regarding the level of wages, however, NUMMI's situation was close to the big three; the average annual earnings for assemblers in NUMMI in 1989 was \$36 013, while that of the Toyota Kentucky plant was \$29 598 (Kenney and Florida, 1993).

In both NUMMI and the big three, the UAW was organized, while in Toyota, Honda and Nissan the UAW was not organized. The UAW, different from Japanese company unions, is organized across the US automotive industry and has the power to launch a strike (see Chapter 7). The unionized production plants in the United States have to persuade the UAW to implement TPS. In the case of NUMMI, as already shown, the union accepted the introduction of work teams and rotation. However, the union did not support the introduction of Japanese payment systems in which a worker's pay can be upgraded in accordance with an increase in experience and skills over time. Nonunionized Japanese transplants in the United States also did not introduce such a Japanese payment system. Generally in Japanese transplants, it took only 18 months for workers to reach the top of the pay scale, in which wage increases accrued equally to all workers across the board (Kenney and Florida, 1993). This means that payment systems in the United States did not help to motivate workers to develop skills and knowledge related to their jobs inside the company, and that there are many difficulties in transferring Japanese HRM to the United States.

We can observe a variety of patterns in the implementation of TPS and Japanese HRM in other countries. According to the author's research on automotive and auto-parts makers conducted in the United Kingdom (Japanese transplants) and Germany (German plants) in 2010–2011, all the German cases implemented suggestion schemes, while only less than one-third of British cases implemented such schemes. Both British and German plants implemented team-based *kaizen* activities. However, the styles in these cases were different from Japanese shopfloor-based activities. Only Toyota UK implemented QC circles that were similar to the Japanese style in which all the shopfloor workers participated. In most of the other cases, project-based activities were conducted, in which only some selected workers and special technical and engineering staff participated. Regarding HRM, job rotation and internal promotion were observed in both British and German cases. However, shopfloor workers could be promoted only to 'Meisters' (supervisors) in most of the German cases, while they could be promoted to executives in more than half of the British cases. Regarding industrial relations, trade unions had strong negotiating power over pay and conditions as well as *kaizen* implementation in Germany, while in the United Kingdom only large plants that had a long history tended to recognize the power of trade unions.

When we look at Japanese transplants in China, according to the author's research on the automotive and auto-parts industries conducted in 2009, most of them looked more actively to try and implement suggestion schemes and QC circles as well as 5S. Particularly, Toyota made a strong effort to implement QC circles and organized an annual QC convention in which all the representative circles from Chinese group plants got together and gave a presentation. Such a convention was also observed in Toyota Europe. Regarding HRM, all the Japanese transplants tried to implement Japanese style payment and promotion systems, in which a worker's wage grade was upgraded in accordance with an increase in experience and skills. As mentioned before, the successful implementation of these systems depends on the practice of long-term employment. However, no case recognized the establishment of such a practice in China, which means employees easily change jobs if they can find a company who provides higher pay and better conditions. In some plants with a long history in China, there were a few Chinese managers who were promoted from within the company.

From the above, we can understand that 'Japanization' of the global automotive industry has progressed. In many countries, team-based *kaizen* activities have been implemented in some form. Furthermore, some automotive firms, such as Volkswagen and Daimler, have used a more synchronized form of parts supply systems than JIT that is sometimes called Just-in Sequence. However, there are many difficulties in transferring Japanese labour relations and HRM to other countries because of existing institutions and practices in each country. On the other hand, Japanese labour relations and HRM themselves have changed, particularly since the Japanese bubble economy burst, which will be described in more detail in the next section.

8.4 Changes in Japanese Labour Relations and HRM

The effectiveness of Japanese HRM began to be questioned around the early 1990s when the Japanese bubble economy burst, which indicated the end of the high economic growth period in Japan. A report titled *Shinjidai no Nihonteki Keiei* (Japanese-style management in the new era) published by Nikkeiren (Japan Federation of Employers' Associations) in 1995 shows that the necessity of transforming Japanese HRM had been clearly recognized by key players in the Japanese economy. This report suggests a possible way forward for Japanese HRM that allows for the coexistence of various forms of employment including not only lifetime employment based on an open contract, but also employment based on a short-term contract. In terms of payment systems, this report also promotes a shift from seniority-based pay to that based on individual ability and performance.

The future direction envisaged by Nikkeiren was actually supported by the ruling Liberal Democratic Party in Japan. This government had actively promoted the relaxation of labour regulations in Japan. Some of the programmes, such as the Revision of Worker Dispatch Law, allowed Japanese companies to easily employ temporary workers. As a result, the ratio of non-regular employees to regular employees that had been around 20% in the early 1990s increased to more than 30% after 2003 (34.4% in 2010). Regarding pay systems, performance-related pay (PRP) was first introduced by Fujitsu in 1993, and then became widely used by Japanese (mainly large) companies. According to a report issued by the Ministry of Health, Labour and Welfare in 2004, more than 80% of large Japanese companies that employed more than 1000 people used PRP in some form. Regarding industrial relations in the post-bubble age, large Japanese companies have maintained their company unions, the membership rate of which decreased, however, from 24.5% in 1990 to 18.5% in 2009.

These changes in Japanese HRM shows that Japanese companies began to introduce more individualistic, market-oriented methods in their HR practices, the main aim of which is to reduce labour costs, thereby increasing competitiveness against emerging economies. Actually, because of this, Toyota has reformed its HR practices. Regarding employment practices, Toyota drastically increased the number of non-regular employees following the Revision of Worker Dispatch Law in 2004; suddenly the number increased to almost 30% in 2006 from just over 10% in 2004. Toyota also implemented reforms to its payment system. As Table 8.3 shows, Toyota stopped using basic pay that was more seniority-based and began to use individual ability-based pay that was more performance-based in 1999. Furthermore, in 2003 Toyota stopped using age-based pay and started to use maturity and role-based pay (Nikkei Newspaper, 2003). Around 2000, Toyota also reformed its job grade system, thereby decreasing the number of grades for skilled workers from 9 to 6.

Toyota's HR reforms described above show that Toyota began to implement more individualistic, performance-based HR practices, which had an impact on the implementation of TPS, particularly OJT that supports such implementation. According to Mr Miyazaki as the director of Toyota's HR department, this reform has led to a situation where 'Toyota's traditional culture in which employees coach each other has decreased', and as a result Toyota started to reevaluate the importance of teamwork, instead of individuals' expertise (Nihon Keizai Shinbun, 2007, 25 December). Regarding the increase in non-regular workers, Toyota decided to allow non-regular workers to become members of Toyota's labour union in 2007. At the same time, Toyota also positively re-employed selected non-regular workers as regular workers (Nihon Keizai Shinbun, 2007, 14 December). Consequently, Toyota decreased the number of its non-regular employees as follows: 26.8% in 2007, 25.4% in 2008, 19.8% in 2009 and 12.2% in 2010.

The author's research on Japanese auto-parts makers conducted in 2010–2011 also shows another impact of HR reforms on the implementation of TPS. This research shows the percentage of operators who were employed on temporary contracts in the Japanese plants was 28.6% on average, compared to 8.7% in the British cases and 7.5% in the German cases. Two out of eight Japanese plants had a workforce of which 60% consisted of temporary workers.

Table 8.3 Reform of payment system for Toyota's skilled workers in 1999

Old		New
Ability-based pay (20%)	⇨	Qualification-based pay (30%)
Productivity-based pay (20%)		Productivity-based pay (20%)
Basic pay (40%)		Individual ability-based pay (30%)
Age-based pay (20%)		Age-based pay (20%)

Note: Compiled from data supplied by Nikkei Industry Newspaper (1999), The Japan Institute for Labour Policy and Training, 'Cases of Personnel and Labour Management' http://www.jil.go.jp/mm/hrm/20010727.html, accessed 14 February 2014, Sugiyama (2004).

The high percentage of temporary workers in Japanese plants sometimes had a negative influence on *kaizen* activities. In some cases, managers recognized that there had been a weakening in workers' capability to conduct quality analysis as well as doing OJT or teaching each other skills. Furthermore, in a Japanese plant where 60% of the workforce consisted of temporary workers, team leaders and skilled workers were busy with rework and helping temporary workers, which deprived them of an opportunity to do *kaizen*. This situation would also deprive them of an opportunity to develop regular workers through OJT. In most of the cases, Japanese managers regarded the revaluation of OJT and teamwork as important future challenges.

8.5 Concluding Remarks

As already mentioned, Japanization has steadily advanced in the global automotive industry. Particularly, team-based *kaizen* activities have already been implemented in some form in many countries. However, there is still a significant difference in the degree of implementation between Japan and other countries. This is partly because of the difficulty to transfer Japanese HRM to other countries, as also outlined in Chapter 5. As mentioned in Section 8.2, collaborative human development by OJT is the key to the successful implementation of TPS, which is greatly supported by Japanese HRM. In the United States and some European countries, their own institutionalized industrial relations have become a barrier to the implementation of Japanese payment systems that play a part in motivating employees. In China, even if it is possible to introduce similar payment systems to the Japanese ones, it is difficult to establish a long-term employment practice in which employees willingly support each other. Furthermore, even in Japan after the bubble burst, there has been a weakening trend in the implementation of team-based kaizen activities and OJT.

Considering the current and future Japanese economic situation in the age of globalization, it would be difficult for Japanese companies to maintain traditional HRM practices in their unchanged form. One of the important future challenges for Japan is to create human development systems on the understanding that there will be a certain percentage of temporary workers and as a result a decrease in employees' loyalty to the company. In order to do this, it would be useful to consider a hybrid form of training systems between Japanese and market-oriented ones, such as a mixed training system between Japanese OJT and Western more standardized, document-oriented systems. One of the keys for surviving the age of globalization is to learn more about other practices and cultures and try to create a hybrid form of practice to suite indigenous institutions and practices in each country. This would be an important challenge not only for Japanese companies but also for companies throughout the world.

References

Abegglen, J. C. (1958), *The Japanese Factory: Aspects of Its Social Organization*, Glencoe: Free Press.

Abernathy, W. J., Clark, K. B. and Kantrow, A. M. (1981), The New Industrial Competition. *Harvard Business Review*, **59**, September–October, 68–81.

Adler, P. S. (1999), Hybridization: Human Resource Management at Two Toyota Plants, in *Remade in America: Transplanting and Transforming Japanese Management Systems*, eds. J. K. Liker, W. M. Fruin and P. S. Adler, New York: Oxford University Press.

Cole, R. E. (1994), Different Quality Paradigms and Their Implications for Organizational Learning, in *The Japanese Firm: Sources of Competitive Strength*, eds. M. Aoki and R. Dore, New York: Oxford University Press, 66–81.

Delbridge, R. (1998), *Life on the Line in Contemporary Manufacturing*, Oxford: Oxford University Press.

Dore, R. (1973), *British Factory, Japanese Factory: The Origins of National Diversity in Industrial Relations*, Berkeley: University of California Press.

Elger, T. and Smith, C. (2005), *Assembling Work: Remaking Factory Regimes in Japanese Multinationals in Britain*, New York: Oxford University Press.

Fucini, J. J. and Fucini, S. (1990), *Working for the Japanese: Inside Mazda's American Auto Plant*, New York: Free Press.

Gankoji, H. (2005), *Toyota roshi management no yushutsu [The exportation of Toyota labour management]*, Kyoto: Minerva Shobo.

Imai, M. (1986), *KAIZEN: The Key to Japan's Competitive Success*, New York: McGraw-Hill.

Imai, M. (1997), *GEMBA KAIZEN: A Commonsense, Low-Cost Approach to Management*, New York: McGraw-Hill.

Ishikawa, K. (1985), *What Is Total Quality Control? The Japanese Way*. London: Prentice-Hall.

JETRO (2014), Japan's Outward FDI by Country/Region. https://www.jetro.go.jp/en/reports/statistics/. Accessed 14 February 2014.

Kenney, M. and Florida, R. (1988), Beyond Mass Production: Production and the Labor Process in Japan, *Politics & Society*, **16**, 121–158.

Kenney, M. and Florida, R. (1993), *Beyond Mass Production: The Japanese System and Its Transfer to the U.S.*, New York: Oxford University Press.

Koike, K. (1994), Learning and Incentive Systems in Japanese Industry, in *The Japanese Firm: Sources of Competitive Strength*, eds. M. Aoki and R. Dore, New York: Oxford University Press, 41–65.

Kumon, H. and Abo, T. (eds.) (2004), *The Hybrid Factory in Europe: The Japanese Management and Production*, Basingstoke: Palgrave Macmillan.

Liker, J. K. (2004), *The Toyota Way: 14 Management Principles from the World's Greatest Manufacturer*, New York: McGraw-Hill.

Monden, Y. (1998) *Toyota Production System: An Integrated Approach to Just-in-Time*, 3rd ed. Norcross, GA: Engineering & Management Press.

Nikkei Industry Newspaper (1999), Toyota nenreiku haishi [The abolition of Toyota's Age-based pay], *Nikkei Sangyo Shimbun [Nikkei Industry Newspaper]*, 8 July 1999.

Nikkei Newspaper (2003) Nenreiku haishi: Ginosholu mo taisho ni [The abolition of Age-based pay in skilled workers as well], *Nikkei Shimbun [Nikkei Newspaper]*, 14 December 2003.

Ohno, T. (1988) *Toyota Production System: Beyond Large-Scale Production*. New York: Productivity Press.

Oliver, N. and Wilkinson, B. (1992), *The Japanization of British industry: New Developments in the 1990s*, 2nd ed. Oxford: Blackwell.

Saruta, M. (2007), *Toyota Way to Jinji-kanri, Roshi-kankei* [The 'Toyota Way' and labour-management relations], Tokyo: Zemukeiri-kyokai.

Schonberger, R. (1982), *Japanese Manufacturing Techniques*, New York: Free Press.

Sugiyama, N. (2004), Toyota no Chingin saido [Toyota's wage system], *Chingin to Shakaihosho* [*Wage and social security*], No. 1371 and 1372.

Womack, J. P., Jones, D. T. and Roos, D. (1990), *The Machine that Changed the World: Based on the Massachusetts Institute of Technology 5-million Dollar 5-Year Study on the Future of the Automobile*, New York: Rawson Associates.

9

The Rise of South Korean (or Korean) Automobile Industry

Seunghwan Ku
Faculty of Business Administration, Kyoto Sangyo University, Kyoto, Japan

9.1 Introduction

In recent years, South Korea's automobile industry has undergone remarkable development and has become one of the largest export industries (joining shipbuilding, semiconductors, steel, electronics and petrochemicals) supporting the Korean economy. In particular, Hyundai Motor Company (HMC) has played a major role in this development. On the heels of the Asian financial crisis in 1997, HMC merged with Kia Motors (Kia) in 1998 and has been increasing production and sales volumes globally. HMC has been expanding primarily in emerging countries with a particular focus on 'Quality Management'. In terms of its 2012 production volume, the company has grown to occupy a place among the top five globally (7 120 000 vehicles) after Toyota, GM, VW and Renault-Nissan. This was a position that was unimaginable 20 years ago.

Nevertheless, South Korea's automobile industry has been viewed as having problems due to a weak supplier system, weak basic technology, militant labour unions and low productivity. How was this Korean automobile company able to achieve its current position under such circumstances? It is said generally that the key reasons behind HMC's success are the depreciation of the Won, success in emerging countries, an aggressive marketing strategy, quality improvement and a design-focused product strategy. However, there seems to be no sufficient explanation of the success of HMC.

Accordingly, in addition to outlining the development process of South Korea's automobile industry including socio-environmental systems, this chapter focuses on the Korean supplier systems of Hyundai and Kia formulated within these processes and attempts to analyse the following: (i) the growth factors for HMC, (ii) the formulation and effect of supplier systems in the South Korean automobile industry from a historical perspective and

The Global Automotive Industry, First Edition. Edited by Paul Nieuwenhuis and Peter Wells.

(iii) the influence that the relationship between labour and management on the production floor has on management and production systems.

9.2 A Brief History of South Korean Automobile Industry and the Performance of HMC

9.2.1 Brief History of South Korean Automobile Industry

South Korea's automobile industry has its roots in the colonial period, specifically in 1944 with the establishment of Keisei Precision & Ind., the forerunner of Kia. However, the industry had its full-fledged beginning with the 5-year economic development plan of 1962 following the period of remodelling and assembly of vehicles financed by the American military after the Korean War. In other words, South Korea's automobile industry developed as a direct result of governmental industrial policy. That process of launching automobile businesses began with a technological collaboration between Japanese, European and North American automobile manufacturers. A compilation of these historical transitions is depicted in Figure 9.1.

Since 1962, Kia Industrial and Shinjin Motors have expanded assembly production based on technological collaboration with Japanese companies. In 1973, Kia Industry started production of three-wheeled vehicles through technological collaboration with Toyo Kogyo (currently Mazda) of Japan, and Shinjin began two-wheeled vehicle production through technological collaboration with Honda and expanded their business acquiring Saenara Auto (Kiotsuka, 1990). And, through technical alliance with Toyota, Shinjin tried to assemble passenger cars, buses and trucks. In addition, HMC entered the automobile business by establishing their Ulsan factory through technical collaboration with Ford of the United Kingdom in 1968. In the same year, Asian Auto Company (Asia) began four-wheeled commercial vehicle production through technological collaboration with Fiat and FFSA. So, from this time, competition among four companies, Shinjin, Hyundai, Kia and Asia began in the South Korean automobile industry. In addition, there was Hadonghan Motor, which was a producer of various special purpose vehicles such as tractors, buses and trucks. After changing its name to Dong-A Motor in 1977, it was taken over by Ssangyong Oil in 1985 and changed its name to Ssangyong Motor in 1986.[1]

Under the policy of developing the heavy chemical industry in the first half of the 1970s, Kia decided to stop production of trucks and instead switched to the production of passenger vehicles, and HMC sold a domestic model of its own, the 'Pony'. This was the beginning of a full-blown transition to mass production. On the other hand, business consolidation took place through mergers between domestic manufacturers such as the acquisition of Asia by Kia. In the latter half of the 1970s, HMC began to export in earnest, and vehicle production exceeded 200 000.

As automobile manufacturers faced management crises due to the economic decline after the second oil shock in the late 1970s, 'Automobile Industry Rationalization Measures' were announced by the government in 1982 and new entry into the automobile manufacturing business was prohibited. Afterwards, HMC signed a joint venture contract with Mitsubishi Motors and focused their energies on building a foundation for new car development and

[1] http://www.smotor.com/kr/company/center/history/index.html (accessed 24 July 2013).

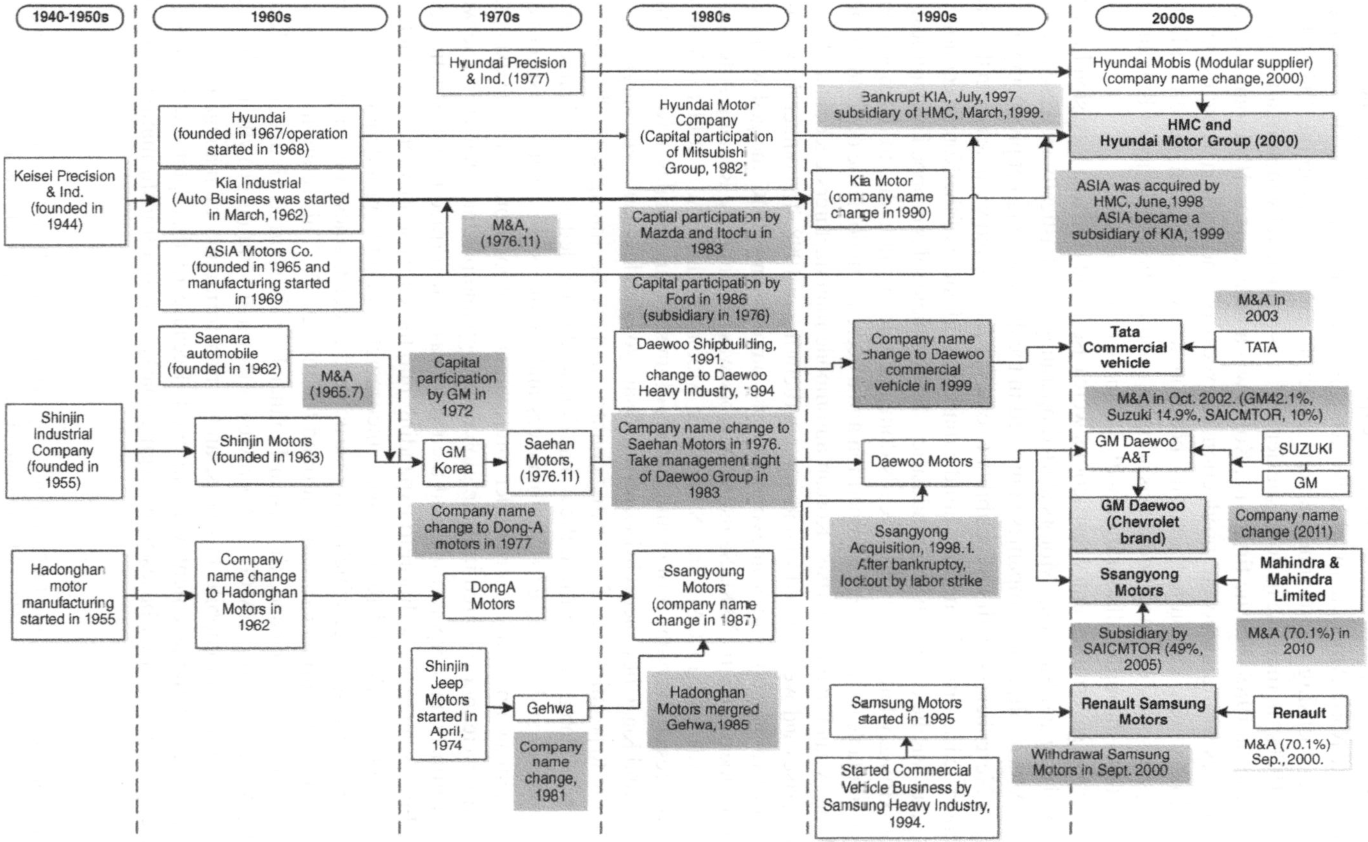

Figure 9.1 Historical restructuring of Korean automobile manufacturers. Source: Compiled from data supplied by Korea Automobile Manufacturers Association (2013:5) and Kiotsuka (1990:77–81)

formulating mass production. Similarly, Daewoo also accelerated new car development. The 1980s represented a phase for establishing an export base, symbolized by the export of Hyundai's Pony 2 to North America. In other words, Korean automobile manufactures adopted low-cost strategies focusing on low-end markets in developed countries in the 1980s. In the latter half of the 1980s, an increase in domestic demand was propelled by motorization spurred on by the Olympics in 1988. For the first time, Korea was among the top 10 automobile manufacturing countries with automobile production in excess of one million vehicles, and the automobile industry was catapulted into its position as Korea's main export industry.[2] Furthermore, Japanese-style management was adopted in production facilities of Korean automobile manufactures through collaboration with Japanese companies.[3]

In the early 1990s, government entry restrictions were eased, and Hyundai Precision Industry, Daewoo Shipbuilding and Samsung Motors started operating and drastically increased production capability. As a result, nine companies competed in the domestic market producing around 1.5 million cars. Nevertheless, with susceptibility to external factors and weaknesses in the structure of the export-led automobile industry amid a marked weakening of the Korean economy in the mid-1990s, as was the case with other industries, 1998 vehicle sales dropped precipitously to 780000, approximately half of the prior year's domestic sales volume due to deterioration in profitability attributable to retail price competition and a decrease in domestic demand. In other words, amid the Asian Financial Crisis, Korean automobile manufacturers encountered extensive economic crises and were forced to undergo restructuring in 1998. To counter these crises, Daewoo Automotive merged with Ssangyong Motor Company and HMC merged with ASIA Motors in December 1997. In 2000, Samsung Motors was acquired by Renault. Afterwards, Daewoo Motors was sold to GM, and Ssangyong Motors business was sold to Shanghai Automotive. As a result, the number of manufacturers of finished automobiles in Korea fell from 9 to 5 (Hyundai and Kia, Chevrolet (GM Daewoo), Ssangyong and Renault-Samsung).

Amidst the rise to prominence of emerging country markets, beginning with China and the movement towards global competition, the Asian Financial Crisis led to major changes in the management policies and strategies of automakers and suppliers. From the perspective of capital, the Korean auto manufacturers under multinational corporations, with the exception of HMC and Kia, were incorporated into the global strategy and production division network umbrellas of multinational corporations, and their role was transformed to a production base in East Asia. For example, in addition to expanding into the Chinese market, manufacturers under the ownership of GM and Renault were incorporated into the production division of labour in East Asia. Ssangyong, which was restructured under Chinese ownership, fulfilled the role as a technology supplier to China around the mid-2000s. This relationship was dissolved recently. On the other hand, the Asian Financial Crisis spurred Hyundai and Kia to form Hyundai Motors Group (HMG) and a review of former volume growth and business strategy focused on the automobile business, and a transformation of management policy towards quality-focused management was initiated. Because of this, plans were made to aggressively expand production bases to emerging markets, and a transformation was made

[2] Imports of passenger cars, which began in 1987, have gradually increased and declined. However, it stayed at a level of less than 1% from 1987 to the mid-1990s. Market share of imported passenger cars has increased up to 9% in 2013(http://www.kaida.co.kr). See Table 9.1.

[3] Lansbury et al. (2007).

Table 9.1 Market share of Hyundai and Kia (%)

	2001	2003	2005	2007	2009	2011	2012	2013
Hyundai	75.7	46.45	47.40	47.15	46.35	40.61	39.93	39.72
Kia		23.09	22.13	20.54	27.23	29.26	28.82	27.51
Imports	NA	1.44	2.57	4.03	4.02	6.23	7.82	9.00

Source: Compiled from data supplied by Korea Automobile Manufacturers Association (2013) and Auto Herald (http://www.autoherald.co.kr/news; 14 June 2013).
Note: Market share including commercial vehicle and imported cars.

towards growth as a global company. As a result, Hyundai and Kia have been able to acquire a dominant domestic market share of about 70% in Korea (Table 9.1).

On the other hand, the effect of the Asian Financial Crisis was not only limited to the consolidation and restructuring of original equipment manufacturer (OEM) makers but also brought restructuring and structural change to the parts industry. The Korean parts industry, which in the early days had relied on imports of parts and materials from Japanese companies, began to slowly develop with an aim towards domestic production. Under conditions that could be called excessive competition in narrow domestic markets, the situation with the supply base was more of an exclusive structure dedicated to a specific manufacturer, and rather than being driven by efficiency, many purchasing processes were influenced by noneconomic factors under the *Chaebol* structure of industrial conglomerates with a broad portfolio of activities. The rise of automakers affiliated with foreign capital caused changes in conventional purchasing practices.

Nevertheless, with the introduction of external capital and automaker restructuring, the expansion of foreign parts manufacturers increased rapidly and transactions were expanded beyond the (Japanese-style) *Keiretsu*, and the conventional closed transaction practice changed. For example, the expansion into Korea by global suppliers such as Delphi, Magna, Bosch and Denso increased. This was evident as the amount of Foreign Direct Investment in the transportation equipment industry for the year 2000 increased to $953 100 000, a 16.7-fold increase compared with 1995.[4] There was large M&A to OEM activity by foreign multinational companies in this period. Furthermore, among the 913 tier 1 suppliers for HMC in 2004 (347 suppliers in 2012), 18% were foreign-owned, and Bosch, Denso, Delphi and Magna accounted for the majority based on total parts sales. In other words, the structure of parts transactions in Korea has transitioned from a focus on domestic suppliers to competition among global suppliers. In order to increase competitiveness from a global perspective, rather than as domestic suppliers within Korea, technological development and resource allocation took place.

The result was a decrease in the relative importance of dedicated suppliers, a change to multiple customer destinations and a rise in the transformation of small-to-medium enterprises to large corporations. Under dramatic environmental changes including development of emerging markets' economy and digital technologies, Korean automobile suppliers could expand their volume of business by learning or imitating advanced technologies. As a result of these suppliers' technological development, Korean automobile companies were able to catch up in

[4] http://www.motic.go.kr/, Foreign Direct Investment Data (Ministry of Trade, Industry & Energy (former Ministry of Knowledge & Economy) (accessed 1 October 2012).

both qualitative and quantitative ways. This type of active use of competitive mechanisms and increased transactions with global suppliers were surely growth factors for HMC.

To summarize, the following are worthy of mention as important factors in the growth process of Korea's automobile industry:

1. Growth through collaboration with Japanese companies such as Mitsubishi, Mazda and Isuzu.
2. The Japanization of work systems and worker control on the production floor.
3. Establishment of an industry structure with an export-oriented strategy and a high level of reliance on exports in order to overcome the limited size of the domestic market.
4. The global dimension of competition and growth of parts suppliers.

9.2.2 The Change in Performance of HMG

As was mentioned previously, the Asian Financial Crisis completely changed HMC's strategy. The company achieved a market position among the global top five through management focus on quality, aggressive overseas expansion and expansion of production capability. In addition, HMC has improved business profit rate from 4.5% in 2006 to 9.1% in 2012 (Table 9.2).

As of 2012, both HMC and Kia accounted for over 70% of the market share of the entire production volume of 4 561 766 vehicles in the domestic market, with HMC producing 1 905 261 vehicles (41%) and Kia producing 1 585 685 vehicles (34%).[5] Furthermore, the companies occupy a monopolistic position with export volume of both Hyundai and Kia at 2 344 084 vehicles, which represents 74% of automobiles exported from Korea.[6] Major export destinations from Korea have been shifting in recent years from North America to Central and South America and to Central and Eastern Africa. The strongest export models are Hyundai's Avante and Sonata and Kia's small cars such as Pride, Morning and Forte as well as sport-utility vehicle (SUVs) such as Tucson ix and the Soul MPV (as of 2012). In particular, Avante has become a best-selling car with cumulative sales of 8.5 million vehicles.

In addition, overseas production and sales volume have been increasing rapidly. After 2000, alongside the rise of emerging markets, HMC and Kia expanded overseas production bases, and production and sales ratios exceeded domestic production and export volume (Table 9.3). The strategic positioning of overseas production bases changed from traditional production supply bases locally to local bases for supplying adjacent markets. In particular, they occupied

Table 9.2 Business profit rate of HMC and Kia (%)

	2006	2007	2008	2009	2010	2011	2012
HMC	4.5	6.4	5.8	7.0	8.8	10.4	9.1
KIA	−0.7	−0.3	1.9	6.2	7.0	8.2	7.5.

Source: Compiled from data supplied by Annual Report and Hankuk Kyongje shinmon (The Korea Economic Daily Newspaper; 26 January 2012).

[5] Korea Automobile Manufacturers Association (2013).

[6] Korea Automobile Manufacturers Association (2013).

Table 9.3 Hyundai Motor Group Overseas manufacturing facilities and production units in 2012

	Overseas manufacturing facilities and capacity	Production (units)
1989	Hyundai Motor Bromont Plant (Closed in 1993)	*Hyundai*
1996	Hyundai India Plant (300 000 units)	India: 638 775
2002	Kia Motor Yancheng Plant (130 000 units)	China: 855 307
2003	Hyundai Motor Beijing Plant (300 000 units)	United States: 361 348
	Hyundai Motor Turkey Plant (100 000 units)	Turkey: 87 008
2005	Hyundai Motor Alabama Plant (300 000 units)	Czech Republic: 303 035
2007	Kia Motor Slovakia Plant (300 000 units)	Russia: 224 420
	Hyundai Motor India 2nd Plant (300 000 units)	Brazil: 27 424
	Kia Motor Yancheng 2nd Plant (300 000 units)	*KIA*
2008	Hyundai Motor Beijing 2nd Plant (300 000 units)	China: 487 580
2009	Hyundai Motor Czech Plant (300 000 units)	Slovakia: 292 050
2010	Kia Motor Georgia Plant (300 000 units)	United States: 358 520
2011	Hyundai Motor Russia Plant (200 000 units)	
2012	Hyundai Motor Brazil Plant (200 000 units)	*Overseas total: 3 665 467*
	Hyundai Motor Beijing 3rd Plant (400 000 units)	

Source: Compiled from data supplied by Hyundai and Kia's Annual Report and Fourin (2009).

market share and market positions relative to Japanese manufacturers in the regions of China and India. It is believed that the tendency for strengthening production bases focusing on overseas markets will further increase due to issues with domestic labour unions, the strengthening of regulations in labour markets and market localization trends.

This increasing trend of sales volume worldwide is affected by not only price competitiveness due to beneficial changes in exchange rates, but significant improvements in quality. There has been a change from quantitative growth to quality-oriented growth or both. To survive global competition and maintain continuous growth, product quality-oriented management was implemented because top management regarded quality as the most important priority.

Quality improvement in HMC was different to the Japanese style where quality improvement within production processes is done by factory workers. HMC's style is that it is quality improvement focusing on final or test line by middle-ranking engineers. In other words, because of labour union issues, HMC adopted a method for quality improvement that did not involve the direct participation of workers in the factory but by middle engineers in quality assurance (this contrasts with Japanese practice as outlined in Chapter 6). The result was an improvement that matched Toyota's quality (Figure 9.2) as clearly shown by J.D. Power initial quality studies. This improvement in quality and increase in overseas sales were related to enhancement of the corporate brand value and resulted in the creation of a virtuous cycle. The brand value of Hyundai increased from \$3.48 billion in 2005 (ranked 84th) to \$6 billion dollars (60th in 2012).[7] Kia also enhanced its brand value to \$4.089 billion (87th in 2012).

[7] http://www.interbrand.com

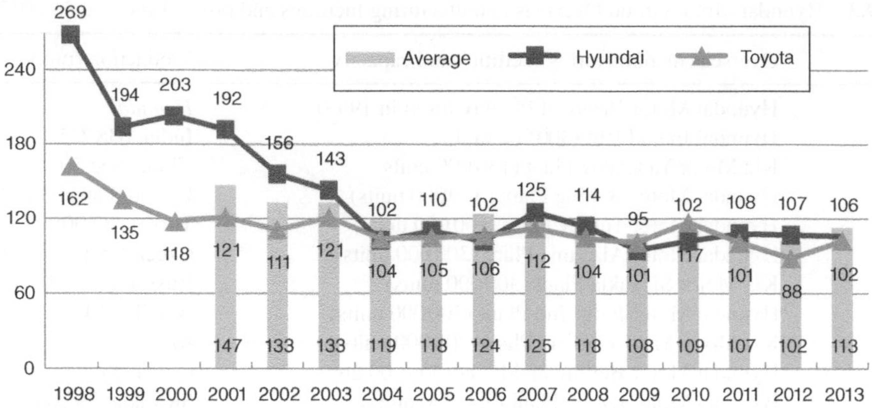

Figure 9.2 US Initial Quality Study results (1998–2013). Source: Compiled from several J.D. Power and Associates reports (1998–2013)

9.3 Considering Five Success Factors of HMC

What exactly are the growth factors of HMC? What did the firm do after the Asian Financial Crisis? The answers lie in Hyundai's and Kia's ability to build a business system that enabled them to share to the utmost degree both firms' synergies in development, production and purchasing. The five success factors are as follows.

9.3.1 Vertical Integration

After merging with Kia in 1998, Hyundai embarked on vertical business integration through the formation of the automotive corporate group (HMG) with automobiles, construction, and iron and steel as the main pillars of HMG. HMG which centred on the three companies of HMC, Kia and Hyundai Mobis embarked on vertical integration through the acquisition of relevant automobile component suppliers and mutual investment in order to procure stable components and materials, internalize core component technology and create mutual synergies. The major mutual investment relationships between firms and HMG are shown in Figure 9.3.

There are two action items of note. The first is the establishment of Hyundai Steel, furnished with three blast furnaces, that creates an integrated production system for automotive steel sheet to reduce the impact of material price fluctuations on profitability. The second is vertical integration of important core functional components suppliers and modular production suppliers to form the automotive corporate group. The formation of HMG became the engine for growth as it resulted in the internalization and propagation of core technologies, the strengthening of purchase bargaining power within markets, the utilization of economies of scale and economies of scope, and the creation of synergy and sharing among groups.

9.3.2 Modularization of Production and Standardization

Around 2003, Hyundai and Kia aggressively adopted modular production. The modular production system was introduced by European and American manufacturers in the 1990s (see also Chapter 5). Modular production in the automobile industry is not about dividing a

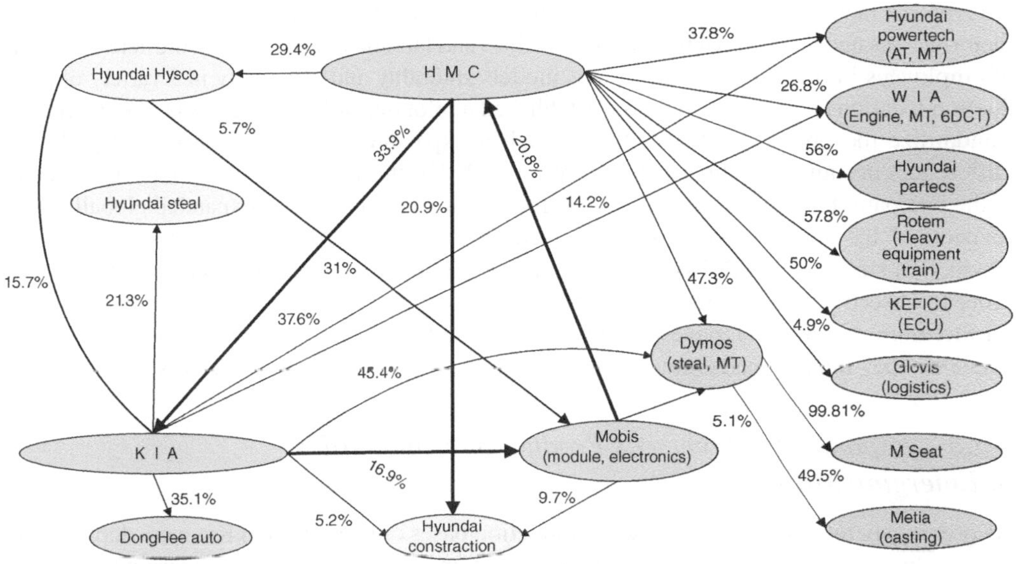

Figure 9.3 Governance structure of major HMG affiliated companies. Note: Based on situation as on 1 April 2012. Source: Compiled from data supplied by Annual reports and Fair Trade Commission (OPNI: http://groupopni.ftc.go.kr)

product into functionally complete units, like with PCs and electric appliances, but refers to a production method to reduce complexity on the main assembly line. That is, modular means to transport and assemble units that are assembled in structurally adjacent component groups by sub-assembly lines or modular suppliers that are distinctly separate from the main assembly line. This production system enables a reduction in the complexity of the main assembly processes in automobile factories. And, when suppliers take on the assembly of the separate work units, OEMs are able to take advantage of the wage differential between suppliers and OEMs. Furthermore, if the units of delivery are modules, supplies are to be delivered after the quality is guaranteed; therefore, automobile manufacturers realize benefits such as (i) reduction in the complexity of the main assembly line and (ii) reduction in man-hours, shorter lead times and decreased scope of supplier control (Ku, 2007).

Hyundai and Kia utilize the cockpit module, front-end module and chassis modules. Internal modular suppliers, such as Mobis and WIA, are located adjacent to final assembly factories and shoulder the responsibility for modular production. In modular production, it is more necessary to share and communicate about order information and operating status between OEMs and suppliers than before. So they have formulated (IT-based) simultaneous production systems. Many of the modular components are sourced from core suppliers within HMG. Therefore, even though this production method is similar to the modular production system adopted by VW, it is different to Japanese manufacturers in terms of in-house manufacture of major modules and component groups.

Up until around 2007, HMG was reliant on foreign suppliers for many functional components or units. However, by utilizing foreign-owned suppliers, HMG has learned the advanced technology and caught up competitors, so they have internalized advanced or applied technology and knowledge.

Overseas factories also have introduced the same production system as used in the domestic factories. It is necessary to control production lines and facilities, skill gaps and the heterogeneity of employees in order to produce various models smoothly and efficiently in a variety of production locations. Therefore, Hyundai established a pilot line at Namyang R&D laboratories to standardize the production line with the goal of expanding the same type of line globally. Differences in equipment and capability are controlled by the production technology department at the company HQ, and the standardized production method has been expanded globally. The revision of the standardized production method is also made by HQ. In other words, the production strategy of HMG is the adoption of a standardized modular production method in order to absorb and reduce the heterogeneity and complexity of various production locations. Furthermore, similar to Japanese companies, to implement the modular production method overseas, they make inroads into foreign markets primarily with group suppliers.

9.3.3 Expansion of Overseas Production Capabilities in Emerging Markets

As previously mentioned, the overseas production bases of Hyundai and Kia were deployed in a way that were commensurate with expansion into China and other emerging markets. Changes of strategic positions in developing countries (from production locations to consumer markets) and the intention to avoid the risk of severe labour problems in domestic markets accelerated the speed of overseas production expansion. As a result, these factors accelerated globalization and localization.

Of note is the magnitude of expansion of production capability. The production capability of overseas factories established by Hyundai and Kia after 2003 reached approximately 300 000 vehicle units. This differs from localization implemented by Japanese manufacturers such as Toyota, namely taking into consideration local standards of pay, the skill level of employees, the environment and so on, to gradually expand production capabilities. We can regard such a strategy as the production strategy with the mindset of larger economies of scale.[8]

9.3.4 Product Strategy

After the merger between Hyundai and Kia, despite preserving each company's retail channel and brand strategy, Hyundai integrated the research laboratories of both firms and established Namyang technical research centres in order to maximize the synergy between the two companies. Accordingly, the product development is based on common platforms and common parts which are developed by the R&D Center. But, product styling which encourages product diversity is separately designed by HMC and Kia. There were 22 platforms in 2002, which were consolidated to 6 platforms by 2013 while the number of car models increased from 29 to 40.[9] In addition to increasing the effects of mass production, this product strategy leads to increased bargaining power. Moreover, the effect was awareness by global suppliers of their

[8] Unlike the global expansion phase of the Daewoo, because the growth period of HMC overlaps the development of digital information technology and the expansion of new emerging market. HMC that has been promoting the product strategy based on quality management was able to get the more effect of economics of scale and economies of scope.
[9] Fourin (2012:126).

desirability as a customer, which served to create an environment that facilitated business with global suppliers who possess advanced technology. As a result, HMC was able to compensate for the weak technological capabilities of domestic suppliers.

It is believed that the success factors of the product strategy of HMC, which takes pride in superior performance in emerging markets, includes the early discovery of needs commensurate with life cycles or standards of living and introduction of sophisticated design models. Particularly, it is the development of local models which began around the start of the twenty-first century represented by the i10, i20 models, which captured the Indian market, and Elantra targeted at the Chinese market.

In addition, their product strategy included the simultaneous introduction of new and old models. For example, HMC did not have an old model strategy which was adopted by competing companies within the Chinese market. HMC and Kia began the simultaneous intro duction of new and old models for the same car from an early stage. Thus, they have garnered success by executing a market sales strategy with products of differing grades. Although this strategy is beginning to change due to changes in current market needs, it must have been a major driving force contributing to the increase in market share to this point.

Although a portion of the test process and test functions, such as road test driving, tuning software development and so on, have been localized, much of the important technology development and design in product development process was conducted by Namyang technical research centres in Korea. Because of the architectural characteristics of the automobile industry, the home country R&D plays a key role in development function to make adjustments with suppliers in the development process and for production prototypes.

9.3.5 Quality Focused and Design Focused Management

As previously mentioned, in order for Hyundai to compete globally, 'Quality Management' was introduced in 1999 at the insistence of the company owners and the General Quality Headquarters, which was established in 2002, began taking a leadership role. In order to ensure market quality, strengthening of the quality check system by mid-level engineers, strengthening of quality management for each module unit and the formation of quality control systems that utilized digital technology were implemented. Furthermore, capacity building and control systems for essential suppliers were redesigned to smoothly conduct simultaneous modular production.

First, a policy of promoting global competition and coordination was adopted based on the open purchasing policy, long-term business relationships based on mutual trust, global sourcing and localization. Hence, in addition to focusing on the ability to propose technology, which basically includes quality and cost, integrated electronic purchasing systems 'VAATZ (Value Advanced Automotive Trade Zone)' based on competition were constructed, and Hyundai and Kia adopted a joint purchasing policy. They also introduced a five-star system for the objective evaluation and publication of supplier quality control systems and quality levels to strengthen their competitiveness through fair competition and monitoring. Furthermore, improvement of technological capability and quality was undertaken by building various support systems including guest engineer programmes, rental of test facilities and so on and through the formation of external support organizations for supporting suppliers. The results of these efforts were the ability to guarantee and maintain quality as demonstrated by the J.D. Power awards.

On the other hand, beginning mid-2000s, Peter Schreyer, designer of the Audi TT, was hired as general manager to focus not only on quality, but also on design. As a result, the product strategy began to focus more on product design to fulfil customer needs.

9.4 Characteristics of HRM in HMC and Effects on the Management System

Militant labour union movements and confrontational labour–management relationships, along with fragmentation and automation of work, and internal competition systems can be thought of as characteristics of Korea's automobile industry in the field of Human Relationship Management.

9.4.1 Militant Trade Union Movement and Confrontational Labour-Management Relations

Korea was able to achieve phenomenal economic growth in a short time through state-led economic policies. However, during Korea's economic growth, it is also true that export-led economic policies centred on *Zaibatsu/Chaebol* (conglomerate) led to major sacrifices and burdens on workers. One example is the militant trade union movements. After 1988, working conditions and wage levels improved significantly during the pro-democracy movements; however, the strength of the labour unions is still the Achilles' heel to the growth of automobile manufacturers even today. Confrontational labour–management relationships are behind this problem. For example, there was a layoff of 10 000 employees by Hyundai during the Asian Financial Crisis and the layoff of 1 750 employees by Daewoo prior to the merger with GM. These types of action combined with the labour market characteristic of rigidity due to the division of the internal labour markets are reasons for the ongoing confrontational relationship between labour and management. Some of the labour unions in large corporations are either 'labor aristocracy' or special interest groups. In the case of Hyundai, the average wage of full-time factory employees is actually higher than that of white-collar management.

Another example is the difference in pay between small- and medium-sized firms and large firms. On the other hand, wage differentials between suppliers have also become significant. Wages are even lower in tier two suppliers. These circumstances have further strengthened modular production utilizing outsourcing. The rigidity of the internal labour market due to wage differentials between large corporations and medium and small firms is a cause for a further strengthening of the militant trade union movement. Accordingly, domestic OEMs have responded to changes in production volumes by increasing ratios of contract-based and temporary staff without increasing full-time employees. However, the operation of factories taking advantage of wage differentials has reached its limits in recent years as regulations for contract-based and temporary workers have been amended.

9.4.2 Fragmentation and Automation of Work

Although methods similar to the Japanese style have been adopted from the perspective of work organization, labour–management relations remain confrontational. Differing from Japan (see Chapter 8), which has professional training systems and career management tied to employment stability, Korea has frictional labour–management relations that makes such programmes difficult to implement. Therefore, a production system focusing on automation that can be called the new-Fordism has been promoted (Choi et al., 2008:97–141). This differs from the Toyota Production System with its focus on humans on the shopfloor.

At many Korean automobile manufacturers, labour unions have annually made extremely strong requests for wage growth and performance-based wages, which makes it difficult to change working hours or to modify production models in accordance with fluctuating market demands. Because of this, the industry is heading towards more standardized and fragmented processes as well as a higher rate of automation due to an environment that does not allow the Japanese way of passing on skills and incorporating improvements and quality that depend on the factory workers. Taking into consideration the perpetual problems with labour unions, tighter regulations for part-time workers, gaps in productivity between overseas locations and domestic locations (Park, 2013), and so on, it is expected that new future domestic investment will be withheld in favour of acceleration on standardization and automation of production. However, HMC will expand and strengthen overseas production bases and standardization and automation in production systems.

9.4.3 Internal Competition Systems

On the one hand, the performance-based principle is employed in personnel evaluations in managerial and technological departments. Therefore, internal competition is fierce which sometimes causes obstacles to interdepartmental collaboration, knowledge and information sharing. Conversely, the strengths attributed to family-owned companies, such as speedy decision-making, decisiveness and the ability to execute, are the driving forces for generating economies of speed by top management. In other words, at Hyundai and Kia, it can be said that speedy decision-making and execution and the internal performance-based competition system complement each other.

9.5 Conclusion: New Challenges for the Korean Auto Makers as Multinational Enterprises

HMC underwent concentrated growth while receiving technological support from overseas companies, and was transformed into a multinational company around the time of the Asian Financial Crisis. However, in addition to issues of improvement in product brand power, diversification of models, stabilization of product development processes and regulation, and so on, the South Korean automobile industry including HMC is exposed to the following issues.

First, there is the problem of domestic labour unions and productivity. HMC's labour unions are one of the most militant unions in Korea (Lansbury et al., 2007). Every year, production troubles occur due to strikes causing obstacles to factory and corporate operations. On the other hand, there is a need to increase productivity and production models, which increases a need for switching in personnel on a production line. Furthermore, productivity is relatively low compared to overseas production bases. Efforts to solve these problems will not only increase productivity, but also will be the key factor for expanding global production.

The second issue is the formulation of organizational capabilities to cope with complexity. As understood by Toyota's recall incidents which occurred at the end of 2009 and start of 2010 in the United States (the essential problem of recall was software bugs), design is the cause of many recalls. In particular, the formulation of organizational capabilities to respond quickly amidst increasingly complex problems of software due to computerization of automobiles has become a life-and-death problem. Software control technology is extremely important as a factor influencing product quality and product capability in the development of hybrids and environmentally friendly vehicles. It will be critical to rapidly establish development capabilities to cope with complex software problems at the stage of product development. To this end, raising the current R&D investment capability has become an urgent need.

The third issue is the formation of networks between foreign production bases and that of organic distribution networks. The integral architecture of automobiles requires the formation of close relationships with suppliers. On this point, the ability to reallocate dynamic resources, in other words, to review the accurate division of labour among regional bases in accordance with the changing market and needs is necessary for the smooth operation of overseas production. Furthermore, in order to overcome constraints, such as geographical distance between bases in developing countries, efficient logistics for finished vehicles and parts is required. In order for this to happen, it will be critical not only to improve the organizational capability of suppliers that also expand overseas but also to locate, evaluate and develop the utilization capacity of local suppliers.

References

Auto Herald (2013) Kia Motors of falling market share: http://www.autoherald.co.kr/news; Accessed 14 June 2013.

Choi, J., Jung, J. and S. Hwang (2008) *HankukKyengjae wa Nodongchejebaenhwa (Korean Economy and the Change of Labor System)*, Seoul: Korea Labor Institute.

Fair Trade Commission (2012) OPNI: http://groupopni.ftc.go.kr. Accessed 27 August 2012.

Fourin (2009) *Kankoku Jidousya and Buhinnsangyou (Korean Automobile and Parts Industry)*. Fourin: Nagoya.

Fourin (2012) *Sekai Jidousya Buhinnsanngyou nennkan (World Auto Parts Industry Yearbook)*. Fourin: Nagoya.

Hyundai Motor Company (2003–2013). Annual Report.

J.D. Power and Associates (1998–2013) U.S. Initial Quality Study Results (IQS): http://www.jdpower.com/automotive. Accessed 27 March 2015.

Kia Motors Corp (2003–2013). Annual Report.

Kiotsuka, M. (1990) *The Hole Aspect of Korean Auto Industry (kannkokujidousysanngyo no subete)*, Tokyo; TED Sogokyouikukikaku.

Korea Automobile Manufacturers Association (2013) *2013 Hankuk Jadongcha sanup, (2013 Korean Automobile Industry)*, Seoul: KAMA. http://www.kama.or.kr/. Accessed 27 March 2015.

Ku, S. (2007) *Seihinn akitekucya no dainamizumu—mojyuruka, chikitoukou, kigyoukannrennkei (Dynamism of Product Architecture: Modulariztion, Knowledge Integration and Inter-firm Linkage)*, Minerva: Tokyo.

Lansbury, R. D., Suh, C. and S. Kwon (2007) *The Global Korean Motor Industry: The Hyundai Motor Company's Global Strategy*, Routledge: New York.

Park, J. (2013) Column: Hyundai Jadongcha Gojang Saengsanseng Noran Jinshin (Column: The Truth of the Productivity in Hyndai Motors's factory is), *HanKyoreh Newspaper*. 21 August 2013.

10

China's Car Industry

Paul Nieuwenhuis[1] and Xiao Lin[2]
[1] *Centre for Automotive Industry Research and Electric Vehicle Centre of Excellence, Cardiff Business School, Cardiff University, Cardiff, Wales, UK*
[2] *Centre for Automotive Industry Research, Logistics and Operations Management Section, Cardiff Business School, Cardiff University, Cardiff, Wales, UK*

10.1 Background

In the first decade of this century, the Chinese were taking to automobility with an eagerness reflecting long pent-up demand (Conover, 2006; Gallagher, 2006), yet modern cars constitute an exceptionally inefficient way to transport one or two people in terms of energy and resource use, as well as roadspace, as Chinese cities are finding to their cost. The existing car paradigm has served the less populated industrial countries reasonably well for a century, but is now recognized even in those countries as ultimately unsustainable, as strategies for its longer term survival or replacement are being sought (Dennis and Urry, 2009; Nieuwenhuis, 2014). The car may benefit some people, but significant sections of the urban population are excluded, as cars and their infrastructure remove urban space and reduce streetlife, limiting social interaction and quality of urban life. Similarly in rural areas, many people are forced to make way for new car-based infrastructure schemes, which primarily benefit wealthier urbanites.

The Tenth Five-Year Plan of the PRC put in place mechanisms allowing an environmental strategy to be developed in conjunction with the Eleventh Five-Year Plan with further developments in the Twelfth Five-Year Plan. However, China – in common with most industrialized and industrializing countries – is still on an unsustainable automotive trajectory. Nevertheless, the Chinese authorities were showing signs of at least a willingness to tackle the problems relatively early on this development trajectory (Nieuwenhuis and Zapata, 2005). Thus even by 2005, the Chinese authorities had already identified a number of possible strategies to reduce the environmental impact of their cars, notably by promoting smaller cars, diesel cars and alternative fuel and powertrain (Nieuwenhuis and Zapata, 2005; Zhengzheng, 2006). The latter have been supported under the aegis of 'new energy' vehicles. Subsequent policy initiatives

The Global Automotive Industry, First Edition. Edited by Paul Nieuwenhuis and Peter Wells.

have further built on these initiatives and display a growing willingness to take on a leadership role in terms of environmental measures. Most recent among these are moves to significantly reduce the country's reliance on coal for electricity generation, something that will make its promotion of electric vehicles (EVs) and the leadership role taken in this area by firms such as BYD, as well as the large electric bicycle market (see Chapter 17), significantly more credible in terms of carbon reduction (New Scientist, 2014).

10.2 Pre-History

It may surprise many that China played a role in the pre-history of the car. Before internal combustion engines powered vehicles, there was steam – an external combustion process. However, there were powered vehicles before steam. In Europe, some clockwork vehicles were developed in the late middle ages, but more spectacular were the two wind-powered vehicles designed by Dutch scientist and mathematician Simon Stevin for his pupil, Prince Maurits of the Netherlands in 1600 (Bouman, 1964; McComb, 1974). The larger one of these could accommodate over 30 people. These land-yachts were inspired by reports from China; Europeans working in Beijing found references to wind-powered vehicles in old Chinese sources. It is likely such vehicles were built in China at one stage, but by the sixteenth century, most of this knowledge had been forgotten. China also played a role in the more significant transition from using direct energy from wind, or human and animal power to steam – fossil fuel. Those same Chinese sources found by the Europeans that yielded the information on wind power also mentioned steam power. This tied in with similar work carried out by the Greek scientist Hero of Alexandria around 2000 years ago. These ideas were developed by a Jesuit priest working in an advisory role at the imperial court in Beijing. Father Verbiest was from the southern Netherlands, an area that later became Belgium. Verbiest used the theoretical information about how to use steam, which he found in the imperial library in Beijing, to develop a small steam-powered vehicle (Golvers and Libbrecht, 1988). The vehicle has not survived, but was apparently used briefly as a novelty gadget to move dishes around the table at imperial banquets. This is generally regarded as the first powered vehicle in the modern sense and certainly as the first steam-powered vehicle (Bentley, 1953; Bouman, 1964; Golvers and Libbrecht, 1988).

Unaware of these Chinese developments, Europeans were also experimenting with steam, although primarily for industrial purposes – initially to pump water from mines. Compared with some Asian countries like China, Europe had always suffered from a relatively low population. This was due to several factors, such as major pandemics – plagues wiped out up to half the population in some parts of Europe in the fifteenth century, though China suffered in a similar manner – and the regular wars between the many small states that made up the continent. Unlike China, Europe has never been successful in unifying itself into a single major political or cultural unit. The Roman empire lasted only a few centuries, while the later attempt by Charlemagne in the eighth century lasted only a generation. A later effort by Napoleon lasted only a few years. Despite the recent formation of the European Union, Europe remains fragmented and diverse.

These low population levels combined with high ambition fuelled by the need for international competition in both war and peace led to early developments of mechanization. Even well before the industrial revolution, Europeans therefore used non-human and non-animal power to do many jobs (Gimpel, 1976). Wind and water mills – some tidal – were in use for various types of industrial activity all over Europe; the use of machines was therefore familiar to Europeans. Hence, when steam was harnessed as a power source combining portability with very high energy density, its adoption was inevitable. Once established as a power source for industrial activities such as pumping water out of mines, powering spinning and weaving machines, sawmills and so on, people started to experiment with powered vehicles. The first such experiment is normally attributed to Frenchman Nicolas Cugnot who built his 'fardier', a steam-powered tractor designed to tow heavy guns, in 1769 (Bentley, 1953; Jacomy and Martin, 1992). The initiative then moved to the United Kingdom where a number of steam-powered coaches were built (Bentley, 1953), tying in well with the other industrial developments in the United Kingdom of what became the industrial revolution.

China developed a quite different attitude to technology; though frequently technology leaders, many such inventions were seen as mere novelties to entertain the leadership. The abundance of labour made machines somewhat redundant as working devices, while the strength of central authority, whether at imperial, provincial or city level, made private initiatives often difficult, unless enjoying official support, as in the case of the ambitious canal-building projects, architecture and, of course, the various great walls. China's early technology leadership position, which could have lead, on paper, to it initiating the industrial revolution, was therefore somewhat dissipated by other factors, not least its comprehensive bureaucracy. So when the car came to China, it came from abroad.

10.3 China's Car Industry

China's car industry started in 1952 was based on imported technologies and was very limited for many years with two large cars aimed at senior party officials, the Shanghai and the Red Flag. Trucks played a more pivotal role in China's centrally planned economy, with Dong Feng's medium duty trucks particularly important. Until the opening up of the economy, cars were restricted to various forms of official use. However, some early joint ventures began to break the monopoly of dated local designs. Chrysler-Jeep (Beijing) and Volkswagen (Shanghai Auto Industry Corporation (SAIC) and FAW) were early entrants in the 1980s, followed by other foreign firms, including General Motors, Iveco, PSA, Toyota-Daihatsu and others (CCFA, 2001). Until recently, Chinese car production was primarily of cars designed abroad. In recent years, this has begun to change, although many designs are still derivative of western or Japanese designs, while others actively engage foreign design and engineering expertise to develop unique products to international standards. What is without doubt, however, is the ability of China's car industry to build cars in high volumes and to increasingly high standards. Nevertheless, exports are still limited, which with the rapidly growing home market is not yet a problem. As the expectations of the home market develop, quality standards are improving, while increasingly, local solutions and designs are emerging. One significant step in this direction is BYD's electric car lead.

The growth in both supply and demand in China over the past decade has been staggering as shown by Figures 10.1 and 10.2, reflecting not only the size of the population, but also the success in spreading a significant proportion of the country's new-found wealth through a large enough segment of the population to create a new car-buying middle class, something other newly emerging economies have not always managed, favouring as they do a small elite

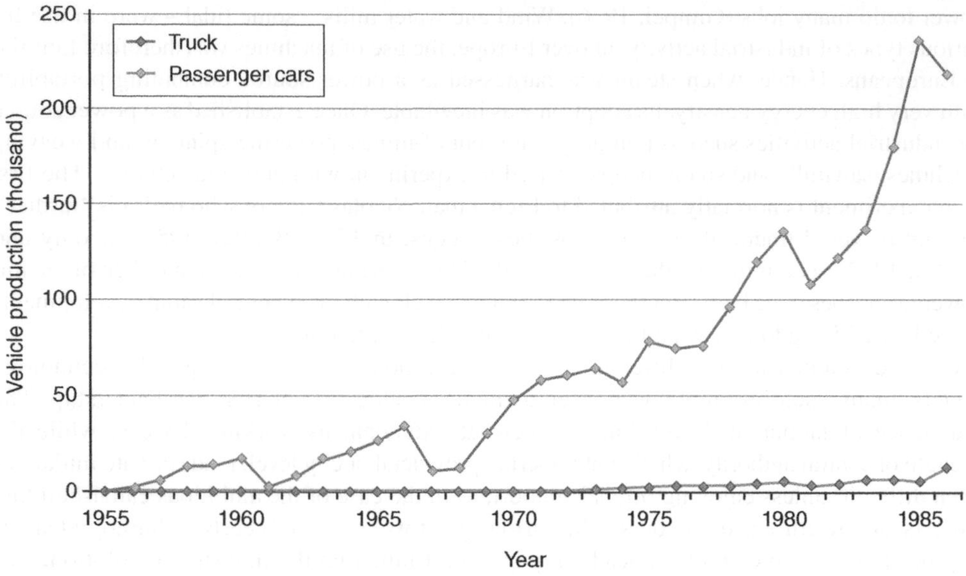

Figure 10.1 China: Vehicle production from 1955 to 1986. Source: Compiled from data in China Automotive Industry Yearbook (1983, 1988) and National Bureau of Statistics of China (1949–1984, 2012). The passenger cars do not include MPV, SUV and cross passenger cars

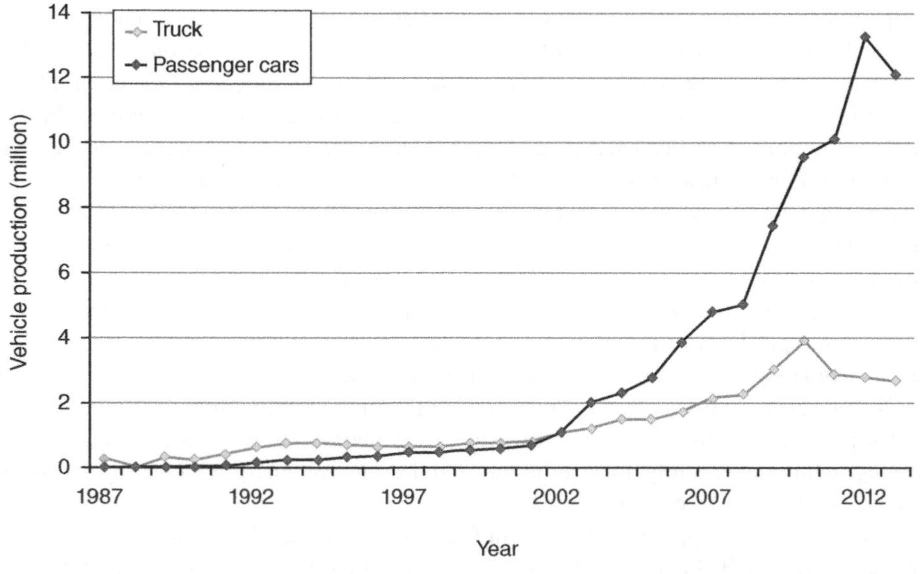

Figure 10.2 Vehicle production from 1986 to 2013. Source: Compiled from, China Automotive Industry Yearbook 1995, 2013; China Auto Industry Development Annual Report 2014; Note: Passenger cars do not include MPV, SUV and cross passenger cars

of wealthy individuals. Nevertheless, most of China's population is still excluded from car ownership, while such car ownership, particularly in urban areas, has created the problems of pollution, congestion, social exclusion and so on outlined in our introduction. In terms of market size and production, China now exceeds the United States and EU, although the extent to which this is economically sustainable is as yet unclear.

An area of weakness is the industry's continuing dependence on foreign technology, which has become embedded in the widespread use of joint ventures. While the success of many of these joint ventures is undeniable, it has the inherent problem of mutual dependence, with the foreign partner dependent on the high sales in China's large home market and the Chinese partner still dependent on foreign design and technology. By contrast, South Korea's car makers, while initially also relying on foreign partnerships, successfully marginalized their foreign partners over time and eventually separated themselves from them. While significant numbers of European and American experts still contribute to Korean products, this is based on Korean-led strategic decisions and in any case reflects the now global nature of Korea's car industry (see Chapter 9).

At present, both sides benefit from the Chinese partnerships, while government policy still makes them largely unavoidable, but as the Chinese market becomes more saturated, tensions may well begin to manifest themselves. This also affects the foreign partners, with VAG and GM particularly vulnerable. Both firms are now very dependent on the Chinese market. In the case of the Volkswagen Group, which in the Chinese market primarily concerns their VW and Audi brands and to a lesser, though growing extent Skoda, China is now their largest single market by some margin. In addition, China has now also become a crucial market for other German firms such as BMW and Mercedes and with the importance of the car industry for the German economy as a whole, it is safe to conclude that Germany's relative success in recent years has been due in no small part to the money made by its car makers in the growth markets of the Far East, particularly China. Any decline in the Chinese economy and the consequent decline in the car market will therefore inevitably have an adverse effect on the German economy and thereby the Eurozone as a whole. Britain's luxury car makers Jaguar-Land Rover, Aston Martin, Bentley and Rolls-Royce are similarly exposed to Far Eastern economies, although they represent a smaller proportion of UK economic activity, while much of this sector is foreign owned with those profits not reinvested in UK-based research and development (R&D) and product development located outside the United Kingdom. As for General Motors, it restructured after the bailout of 2009 primarily around its North American and Pacific hubs, with the latter focusing particularly on China and its partnership with SAIC; in fact, serious attempts were made to dispose of its European division, Opel-Vauxhall. The extent to which China informed that restructuring is also illustrated by the decision to retain the Buick brand, which, though traditionally strong in China, was effectively increasingly marginal in the markets of North America, where now defunct Oldsmobile was slightly stronger. A North America-focused strategy would therefore have seen the demise of Buick rather than Oldsmobile.

It is clear from Table 10.1 that China now represents around a third of global sales for GM, with the US home market relegated to second place with 2.8 million sales in 2013 (GM, 2014). With the increasing inter-dependence of the US and Chinese economies, this clearly makes the group now very vulnerable to any economic fluctuations in either market.

For VAG, the Chinese market share is similar in scale to that in Western Europe, which amounted to 3.65 million in 2013, compared with 3.27 million in China, including imports (Table 10.2). But Europe features a range of markets, each with their own economic dynamics.

Table 10.1 GM's dependence on China

Firm	Sales 2013	Sales 2012
SAIC GM Wuling	1 584 920	1 458 188
Shanghai GM	1 512 000	1 363 532
FAW-GM light CV	59 092	54 847
Total China	3 156 012	2 876 567
GM Global Sales	9 714 652	9 288 277
China share of global sales	32.48	30.96

Source: Compiled from GM, China Auto Market Almanac (2013) and National Bureau of Statistics, China (2012).

Table 10.2 VAG dependence on China

Firm	Sales 2013	Sales 2012
Shanghai VW	1 431 626	1 280 008
FAW-VW	1 607 000	1 328 888
Total China	3 038 626	2 608 896
VAG Global Sales	9 714 652	9 254 742
China share of global sales	31.27	28.19

Source: Compiled from VW, China Auto Market Almanac (2013) and National Bureau of Statistics, China, China Auto Web and other sources.

The German home market, for example, represents around 1.16 million of the European total. Although such statistics provide a snapshot of the industry and these firms in particular, it does show their growing reliance on the Chinese market, such that GM and VAG are as much Chinese firms as they are, respectively, American and German firms. For VAG, the VW brand itself is particularly reliant on the Chinese market, where in 2013, for example, it sold 2 395 696 cars (China Auto Web, 2014).

10.4 The Role of Government

A prominent feature of the automobile industry in China is that it is significantly supported, directed and intervened in by the central government, which leads to both positive and negative results.

10.4.1 Traditional Automobile Industries

The automobile industry in China started from 1952, 3 years after the establishment of the People's Republic of China. Although falling far behind the more automotively developed countries, the automobile sector was subject to direct central government control from the First Five-Year plan onwards. At that stage, the government naturally mainly promoted the development of heavy-duty vehicles, considering the demand for bulk freight transport in a centrally planned economy. In 1979, a government department, the National Mechanical Industry Committee (NMIC), was organized to control oversight of the automobile industry directly. NMIC established a state-owned company, which monopolized the whole supply chain of automobiles in China. An immediate negative result of this exclusive monopoly was that the total industry output in 1986 decreased by 15.59% from 1985, and the profit decreased by 59% (China Automotive Industry Yearbook, 1988).

In 1986, the automobile industry was officially confirmed as a *pillar* industry of China in the Seventh Five-Year Plan (1986–1990). The direction of its development was also changed to the passenger vehicle sector. However, as the automobile industry only focused on truck development for a long period, the passenger vehicle sector was lacking in technologies, manufacturing facilities and skilled labour (see Figure 10.1). The central government decided to initiate a large-scale programme of international cooperation to introduce the much needed investments and technologies through joint ventures (Chu, 2011). Beijing Jeep was the first joint venture, set up by Beijing Automobile Works (BAW) and American Motors (AMC), with a registered capital of $51.03 million. BAW provided manufacturing plant, facilities and $6.6 million investment, and held 68.65% of the total share capital. On the other hand, AMC provided Jeep-branded products, technologies and $8 million cash investment, accounting for 31.35% of the share capital (Zhang, 1995). SAIC also established a joint venture with Volkswagen in 1983, and Guangzhou with Peugeot in 1985, followed by more joint ventures (see Table 10.3).

In 1988, the government proposed a strategy to support six companies, FAW, SAW, SAIC, Beijing Jeep, Tianjin FAW Xiali (TJFAW) and Guangzhou Peugeot, which are all stated-owned. The strategy implied that the government neglected the importance of private companies. The neglect was further strengthened by 'Measures for the Implementation of national automotive industrial policy', which clearly stated that the automobile industry must be controlled by the state (China Automotive Industry Yearbook, 1991). The policy also formulated the automobile industry structure in China. First, it specified the products which should be given priority, such as passenger cars (middle size and below), Iveco series and Isuzu series (light duty vehicles), FAW Jiefang series and Dongfeng series (medium duty vehicles), and Steyr series (heavy duty vehicles). It is noted here that the aforementioned vehicle products mainly relied on joint ventures. Secondly, to protect the domestic manufacturers, the

Table 10.3 Principal joint ventures in China's automotive industry

	Total investment (million dollars)	Investment ratio (%)		Signing date	Duration (year)
		Foreign	China		
Beijing Jeep	51.03	31.35	68.65	1983	20
Shanghai Volkswagen	180	50	50	1983	25
FAW-Volkswagen	791	40	60	1990	25
Guangzhou Peugeot	63.9	34	66	1985	20
Dongfeng Peugeot Citroën Automobile	1261	30	70	1990	20

Source: Compiled from data in Xie and Wu (1997).

government set up high tariffs to restrict the import of passenger cars. Thirdly, the entry requirements for manufactures to the industry were significantly raised (see Table 10.4).

The tariffs were later reduced by the central government to attract more foreign investors, when the 'Automobile Industry Policy' was issued in 1994. Consequently, more joint ventures were established, such as SAIC-GM (1997), Guangzhou-Honda (1998) and Tianjin FAW-Toyota (2000). This policy led to a rapid expansion of the industry's scale in a very short time (Figure 10.2). However, the entry of domestic private companies was still prohibited, with the high standard in the production capacity, passenger cars (150 000 units), light vehicles (100 000 units) and engines (10 000 units) (National Development and Reform Commission, 1994). Under the protection of the government, the stated-owned companies enjoyed high profits, but did not develop the indigenous key technologies and successful brands, losing competitive power in the market. For example, SVW Santana was first launched in 1987 in China and still enjoyed a 48% market share after 10 years (Liu, 1999). The key components, such as engines and transmission, mainly relied on imports (Figure 10.3).

Table 10.4 Entry requirements of automobile manufacture

	Passenger cars	Light vehicles	Engines	Estate cars
Production capacity	300 000	120 000	300 000	30 000
Registered capital (billion Yuan)	2	1	1	0.3
Fixed assets (billion Yuan)	5	2	3	0.5
Technical personnel ratio (%)	18	12	20	10

Source: Compiled from data in China Automotive Industry Yearbook (1991).

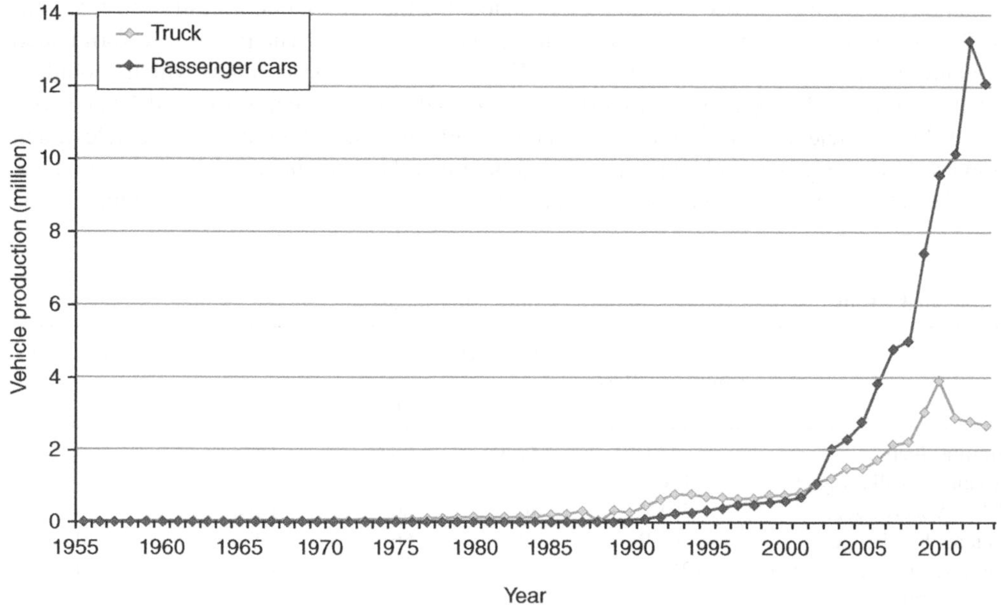

Figure 10.3 Vehicle production from 1955 to 2013

In 2001, China became a member of the World Trade Organization (WTO). With the new pressure from WTO to open the automobile market, the central government gradually relaxed entry restrictions to domestic private companies, granting vehicle assembly permission to four indigenous firms: Chery, Geely, Hafei and Brilliance. In 2003, battery maker BYD also started car production and entered the market. On the other hand, the creation of joint ventures did not cease, as Changan-Ford (2001), Beijing-Hyundai (2002), Brilliance –BMW (2002) and Dongfeng-Nissan (2002) were established. In 2004, the central government issued a second edition of 'Policy on Development of Automobile Industry' (National Development and Reform Commission, 2004), which:

1. removed industry performance requirements (foreign exchange balance, localization rate and export performance) and
2. reduced the administrative approval procedures.

This policy achieved greater progress in terms of the development of domestic private enterprises, leading to a rapid increase in indigenous brands (see Table 10.5). However, the policies still placed key importance on technology imports, mass production and protection of the stated-owned companies and joint ventures. Hence, the R&D sector only accounted for 1.5% of total investment. In 2012, the total R&D investment of all the domestic firms just reached 40.36991 million Yuan (6.5 billion dollars), which was significantly lower than Volkwagen's R&D (11 billion dollars) (Hill et al., 2014).

To sum up, the automobile industry achieved a fast expansion with the great support of the central government, as shown in Figure 10.4. However, the government intervention caused the following unfavourable results:

1. The strategy of 'exchange market for technology' did not achieve the key technology introduction; without giving priority to indigenous R&D.
2. The government overprotected the state-owned companies, whereas seriously restricting private companies. The lack of competition made state-owned companies not motivated to develop the technology.

Table 10.5 Market share of indigenous cars

	Sales of passenger cars (10 000)	
	Indigenous brands	Share of indigenous brands (%)
2004	49.6	19.98
2005	74.1	23.77
2006	115.3	26.80
2007	157	29.5
2008	159	28
2009	259	30
2010	370	32.7
2011	384	31.2
2012	417	31.5

Source: Compiled from, China Automotive Industry Yearbook (2007) and Wen and Li (2013).

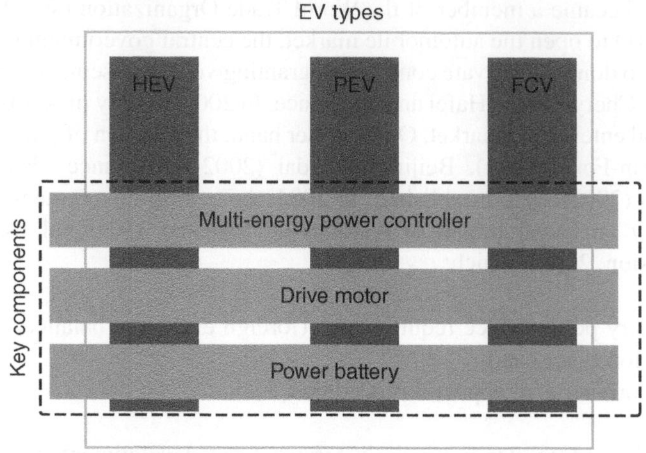

Figure 10.4 The framework on the EV R&D. FCV, fuel-cell vehicle

10.5 New Energy Vehicles

'New energy vehicles' (NEVs) – vehicles using alternative fuels or powertrain – are regarded in many countries as a promising solution to the environmental pollution and energy crisis, and they are also central to the Chinese government's plan for the automobile industry. In this section, we review and analyse these NEV industrial policies, which were intended to promote the industry development from zero to industrialization in 20 years.

10.5.1 R&D Support

R&D in NEVs was mainly directly supported by the central government, which also determined the research direction. This support policy of the central government on R&D can be divided into the following three steps:

1. The Tenth Five-Year Plan (2001–2005).

   The national policy on NEVs started from 2001, when the 'National High Technology Research and Development Program (863 Program) – electric vehicle major project' was launched in the Tenth Five-Year Plan (Xu, 2001). The 863 Program laid a foundation and specified the framework for NEV research. According to the Program, research should focus on three types of EVs: (i) pure electric vehicles (PEVs), (ii) hybrid electric vehicles (HEV) and (iii) fuel cell electric vehicles (FCEVs), and three types of key components: (i) multi-energy powertrain controller, (ii) drive motor and (iii) power battery, which are the critical techniques common in all three types of EVs (see Figure 10.4).

   At the preliminary research stage of EV development, the emphasis was placed on *vehicle assembly*. In this period, 2.38 billion Yuan was invested into R&D EV project, where the vehicle assembly occupied 39%, followed by basic general vehicle technology (17%), powertrain platform (27%) and other key components (17%), as shown in Figure 10.5.

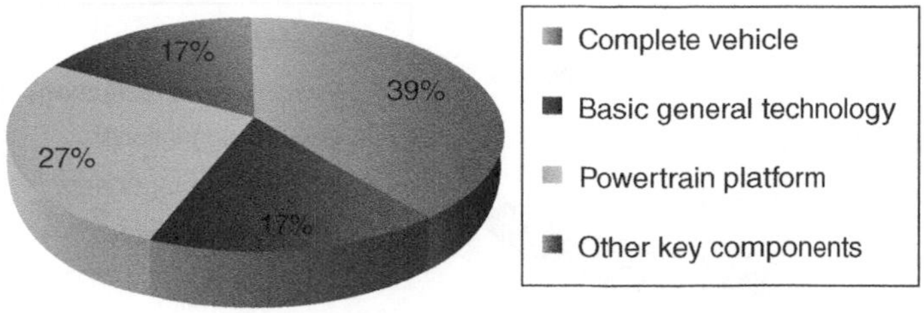

Figure 10.5 The investment in EV research (Tenth Five-Year Plan). Source: Compiled from data in Zheng (2010)

During the Tenth Five-Year Plan, EVs did not step into the commercial production stage, but the government's policy still had a substantial impact on manufacturers. Dongfeng and BYD invested to establish EV R&D centres (Guo and Shi, 2010; Huang,2012).

2. The Eleventh Five-Year Plan (2006–2010)

   During the Eleventh Five-Year Plan (2006–2010), the R&D policy still held the same framework, but the emphasis was switched to the technologies of *key components*. The investment in EV R&D projects increased to 11.1 billion Yuan and the fund was divided into five parts. The key components part took 34% of the total funds, mainly used in the battery sector. The powertrain platform followed up with 31% share while supporting platform took 16% of the funds, which was not invested during the Tenth Five-Year Plan period (see Figure 10.6).

3. The Twelfth Five-Year Plan (2011–2015)

   The new features in the plan included the following:

   (a) Specifying that the strategy should focus on PEVs.
   (b) Incorporating into the main research areas three supporting platforms: standard assessment, resource supply and assembly norms.

10.5.2 Industrialization

The Chinese government also made great effort to industrialize EVs. In December 2007, 'Catalogue for Guiding Industry Restructuring (2007 Version)' was released to guide the related investment, project management, taxation, import/export policies and so on. In March 2009, a detailed plan for developing a NEV industry from 2009 to 2011 was proposed in the 'Automobile Industry Restructuring and Revitalization Plan' issued by the State Council. According to the plan, NEV production would reach 0.5 million in 2011, the sales of NEVs would account for 5% of total sales of passenger cars and the battery production capacity should achieve10 billion AH (The State Council, 2009). In June 2009, 'Rules on the Production of New Energy Vehicles' was released to provide a standard for the assessment of the capacity, manufactures, product safety and quality. It also stipulated the NEV production enterprise qualification and production admission requirements. In the following 3 months, nearly 40 enterprises obtained the qualifications, and more than 100 NEV products were approved (Chen, 2012).

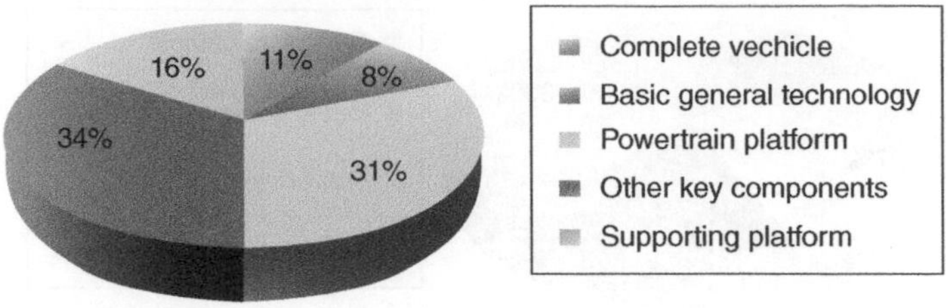

Figure 10.6 The investment in EV research (Eleventh Five-Year Plan). Source: Compiled from data in Zheng (2010)

With these stimulus policies, China Association of Automobile Manufactures convened a meeting (T10 Summit) to discuss NEV development. Thus SAIC, Dongfeng, GAC, BAIC, Brilliance, Chery, JAC and other top 10 automobile manufactures signed the 'New Energy Development Joint Action' with the agreement to jointly and positively become involved in NEV development (Ye, 2010). In addition, the oil companies also entered the NEV-related industry in 2009. China National Offshore Oil invested five billion Yuan in Tianjin Lishen Battery Company, an EV lithium battery manufacturer (Zhang, 2009). SINOPEC and BAIC built a partnership for vehicle fuel, chemical products and NEV development (Guo, 2010). PetroChina and Huaneng Power signed a cooperation agreement on natural gas and power-generation projects.

In June 2012, a specific plan for EVs, the 'Energy-saving and New Energy Automobile Industry Development Plan (2012–2020)', was issued by the The State Council (2012). The plan proposed clear targets for NEV production and sales: (i) cumulative production and sales of EVs would reach 0.5 million in 2015, and (ii) the production would reach two million and cumulative sale five million in 2020. The targets greatly enhanced business confidence. Numerous domestic vehicle companies announced their NEV production and sales targets. For example, FAW Group planned to reach 11 000 production and sales of NEVs, DFG promised to produce and sell 50 000 EVs in 2015, Changan Group aimed to increase the production and sales of NEVs to 150 000 in 2014, and SAIC proposed to increase the NEV production capacity to 300 000 in 2015. Although there is still a whiff of centrally planned economy thinking about such ambitious production targets, in addition, the policy also encouraged the foreign manufactures in China. Honda added four more HEV products in 2013, and Lexus CT 200h was officially launched as the fourth HEV model in China.

To promote the international collaboration in the NEV sector, 'Guidance on the promotion of international development of strategic emerging industries' was issued in 2011. The aim of the Guidance is to encourage Chinese NEV enterprises to participate in international cooperation and competition, including transnational corporations, international standard-setting, investment in overseas factories and so on. In December 2011, the 'Catalogue for the Guidance of Foreign Investment Industries' was released to attract investment in the new energy industry, where many sections involved in the NEV industry are classified into the category in which foreign investment can enjoy political support, including manufacture of fuel cells and mixed fuel powertrain, and other key components of NEVs (National Development and Reform Commission and Ministry of Commerce, 2011).

10.6 Bringing NEVs to Market

10.6.1 Demonstration and Pilot Projects: Strategic Niche Management

In addition to these supply-side measures, sales of EVs were also promoted by the central government, but mainly in the public transport sector through government procurement. In 2001, Beijing, Wuhan, Tianjin, Zhuzhou, Weihai and Hangzhou were specified as EV pilot cities. In 2009, a larger scale of EV demonstration and pilot project was launched, the '10 cities and 1000 units' project. The '10 cities' referred to developing nearly 10 pilot cities each year from 2009 to 2012, and the '1000 units' meant that every pilot city must adopt 1000 EVs in public service such as public transport, taxi, civil service, postal service and so on. It is the first time that the central government detailed the targets of application of NEVs. In order to support the promotion of NEVs in pilot cities, the central government provided 10 billion Yuan financial subsidies. Thirteen cities were selected as the first pilot city group, followed by seven cities in the second pilot city group and five cities in the third group (see Table 10.6).

Among the pilot cities, Shenzhen is the most successful; by the end of 2009, Shenzhen had opened 12 NEV demonstration bus lines, while 244 hybrid electric buses, 20 dual-mode plug-in HEVs and 50 electric taxis have also been put into operation. In addition, 154 charging points, 11 electric bus charging stations and three fast charging stations were established. Furthermore, the first electric taxi service company was established by BYD and Shenzhen Bus Company. Later, Shenzhen invested two billion Yuan to establish the charging points and charging stations, including electric bus charging facilities (0.61 billion Yuan) and public charging facilities (1.39 billion Yuan). The Shenzhen government adopted a franchise strategy to actively lead social capital to establish the facilities. Guangdong Power Grid and China Putian also cooperated to fund the infrastructure construction for obtaining the franchise (Wang et al., 2012).

However, in 2012, the last year of the '10 cities 1000 vehicles' electric vehicle demonstration project implementation, many cities did not reach their initial targets, which meant a serious financial burden on the local governments involved (Zhao and Ao, 2013). A new round of NEV demonstration and pilot projects in 23 pilot cities started in November 2013. According to this plan, in 2015, the cumulative production of NEVs should not be less than 10 000 in the mega-size cities, and other smaller cities or regions should reach the target of 5000 cumulative

Table 10.6 Pilot cities of '10 cities, 1000 units' project

	Pilot cities
The first group	Beijing, Shanghai, Chongqing, Changchun, Dalian, Jinan, Wuhan, Shenzhen, Hefei, Changsha, Kunming and Nanchang
The second group	Tianjin, Guangzhou, Haikou, Zhengzhou, Xiamen, Suzhou and Tangshan
The third group	Shenyang, Chengdu, Nantong, Hohhot and Xiangfan

Source: Compiled from data in Wang et al. (2012).

registrations. To break local protectionism, it is required that the pilot cities and regions cannot set barriers to restrict the purchases of non-local vehicles and must purchase at least 30% non-local vehicles of the total NEVs (Figure 10.7).

10.6.2 Financial Incentives

In 2009, a detailed financial subsidy policy 'Interim Measures on Finance Subsidy of Energy-saving and New energy vehicle Demonstration and Extension' was issued. The policy only applied to the public service sector of pilot cities. According to this financial subsidy policy,

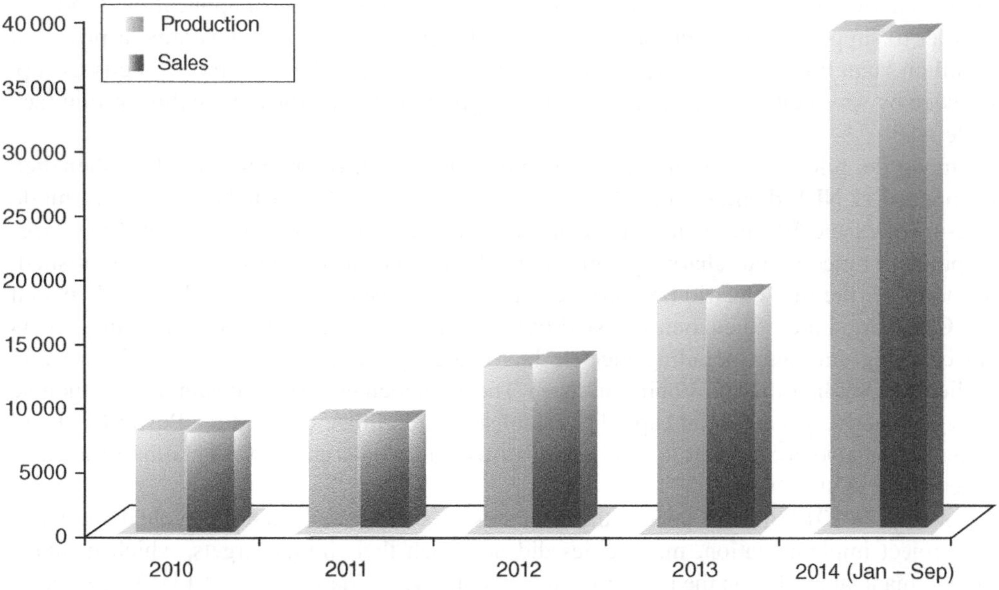

Figure 10.7 The production and sale of EVs in China. Sources: Compiled from: China Automotive Industry Yearbook (2011, 2012); China Auto Industry Development Annual Report (2014)

Table 10.7 Subsidy of passenger cars and light commercial vehicles (2009)

Vehicle	Fuel saving ratio (%)	Maximum wattage ratio			
		BSG vehicles	10–20%	20–30%	30–100%
HEVs	5–10	4	–	–	–
	10–20		28	32	–
	20–30	–	32	36	42
	30–40	–	–	42	45
	40 or above	–	–	–	50
BEVs	100	–	–	–	60
FCVs	100	–	–	–	250

Source: Compiled from Ministry of Finance (2009).
BEV, battery-electric vehicle; BSG vehicles, belt-driven starter generator; FCV, fuel-cell vehicle; HEV, hybrid electric vehicles.

the central government provided the subsidy, with the amount based on the fuel-saving ratio of the NEV (see Table 10.7). In addition, in 2014, BEVs, plug-in HEVs and FCEVs are exempted from vehicle purchase tax, which is usually 8.5% of the total vehicle price from 1 September 2014 to the end of 2017.

Private companies also enjoyed financial subsidies, according to 'Notice on Finance Subsidy of Private Purchase of Energy-saving and New energy vehicles at pilot cities' in May 2010 (Ministry of Finance, 2010). This is a great improvement compared to the restrictive policy in traditional automobile sectors. Table 10.8 shows the subsidy policy for individual purchasing issued in 2013. The financial incentive only applies to the vehicles listed in the 'Catalogue on Recommended Model of Energy-saving and New Energy Vehicle of Demonstration and Pilot Project'. The policy determined the finance subsidy standard by driving range on pure battery power.

The local governments also proposed various incentives to accelerate NEV sales. Take Shanghai as an example. According to the policy updated in 2014 by the Shanghai government, the consumer obtained 40 000 Yuan per unit for BEV purchase and 30 000 Yuan per unit for plug-in hybrid vehicles in Shanghai. In addition, if the passenger vehicle was for private use,

Table 10.8 Financial subsidy standards for new energy vehicles (passenger cars) 1000 Yuan/unit

Vehicle type	Policy year	Driving range R (km) powered by battery			
		$80 \leq R < 150$	$150 \leq R < 250$	$R \geq 250$	$R \geq 50$
BEVs	2014	33.25	47.5	57	/
HEVs	2014	/	/	/	33.25

Source: Compiled from Ministry of Finance (2013) data.

Table 10.9 Charging station and charging point establishment plan

City	2013		2015 (Beijing 2017) plan	
	Charging stations	Charging points	Charging stations	Charging points
Beijing	69	1347	107	188 000
Shanghai	12	1460	50	5 000
Shenzhen	81	3000	218	39 526
Hangzhou	62	620	42	3 500
Tianjin	7	471	66	6 700
Chongqing	11	200	16	275
Hefei	8	2000	6	5 000
Lanzhou	6	82	21	6 650
Taiyuan	7	300	11	1 750
Chengdu	14	880	16	3 000
Xian	5	60	32	480
Wuhu	2	461	10	5 110
Kunming	1	150	2	300

Source: Compiled from data in Xiao (2014) and Wen (2014).

the consumer will get a free special private car license (the average price of private car license is 70 000 Yuan in Shanghai) (Shanghai Government, 2014). In addition, the pilot cities provided appropriate subsidies to support the related infrastructure construction and maintenance. This development can be seen in Table 10.9.

10.7 Conclusions

Within the first decade of the twenty-first century, China had become one of the top three vehicle markets and one of the top three vehicle producers in the world, in some years moving into the number one spot on both counts, leaving more established car regions such as Europe, Japan and North America far behind. However, this growth has come at a cost and much of this cost has been environmental. At the same time, relatively strong central control has allowed rapid and relatively effective action to be taken to limit some of the worst impacts of automobility, with recent limits on emissions, fuel consumption and the promotion of 'new energy' vehicles particularly promising. In particular, the central government has had a positive effect on the development of EVs, especially considering that the sector worldwide is still in a preliminary stage. Compared to the traditional automobile sector, the government released more pro-active policies on R&D, and provided opportunities to private companies. This policy has already paid off with BYD electric taxis and buses now exported to several developed countries.

However, the market for EVs is still not well developed and it is not sustainable to rely solely on government financial support, so perhaps different targets could be pursued in the future such as promoting more private sales of EVs. At the same time, the roll-out of EVs should now be expanded from pilot cities to a wider number of regions.

References

Bentley, J. (1953) *Oldtime Steam Cars, Second Printing 1969*, New York: ARCO.

Bouman, J. (1964) *Oude Auto's en hun Makers*, Bussum: van Dishoeck.

CCFA (2001) *Répertoire mondial des activités de production et d'assemblage de véhicules automobiles*, Paris: Comité des Constructeurs Français d'Automobiles.

Chen, L. (2012) 中国新能源汽车政策盘点 (A review of China new energy vehicle policies), *Auto Industry Research*, **3**, 14–21.

China Auto Market Almanac Editorial Office (2013) *China Auto Market Almanac*, Beijing: China Commercial Publishing House.

China Automotive Technology and Research Centre (1983, 1988, 1991, 1995, 2007, 2011, 2012 and 2013) *China Automotive Industry Yearbook*, Beijing: China Association of Automobile Manufacturers.

China Automotive Technology and Research Centre (2014) *China Auto Industry Development Annual Report*, Shanghai: Shanghai Science and Technology Press.

China Auto Web (2014) http://www.chinaautoweb.com/2014/01/2013-passenger-vehicle-sales-by-brand/. Accessed 16 October 2014.

Chu, W. (2011) How the Chinese government promoted a global automobile industry, *Industrial and Corporate Change*, **20**, 1–42.

Conover, E. (2006) Capitalist Roaders, *The New York Times*, 2 July. http://www.nytimes.com/magazine/02china.html. Accessed 27 March 2015.

Dennis, K. and Urry, J. (2009) *After the Car*, Cambridge: Polity.

Gallagher, K. S. (2006) Limits to Leapfrogging in energy technologies? Evidence from the Chinese automobile industry, *Energy Policy*, **34**, 383–394.

Gimpel, J. (1976), *The Medieval Machine: The Industrial Revolution of the Middle Ages*, London: Pimlico.

GM (2014) General Motors sales up 4 percent in2013. http://gmauthority.com/blog/2014/01/general-motors-sales-up-4-percent-in-2013-by-the-numbers/#ixzz3GJtZ1z3t. Accessed 16 October 2014.

Golvers, N. and Libbrecht, U. (1988) *Astronoom van de Keizer; Ferdinand Verbiest en zijn Europese Sterrenkunde*, Leuven: Davidsfonds.

Guo, J. (2010) 北汽福田与中石油开启深度合作之旅 (The initiation of the cooperation between SINOPEC and BAIC), *Beijing Development and Construction News*, 1 December 2010.

Guo, Y. and Shi, H. (2010) 比亚迪新能源汽车开发中的创新方式研究 (A study on the innovation system of new energy vehicle in BYD), *Journal of Management Case Studies*, **3**(6), 469–478.

Hill, K., Menk, D., Swiechi, B. and Greegger, J. (2014) Just How High-Tech is the Automotive Industry? http://www.cargroup.org/?module=Publications&event=View&pubID=103. Accessed 27 March 2015.

Huang, Z. (2012) 东风汽车稳步推进新能源汽车战略 (The strategy of Dongfeng in developing new energy vehicle), *Auto Review*, **5**, 72–75.

Jacomy, B. and Martin, A.-C. (1992) *Le chariot à feu de M. Cugnot*, Paris: CNAM.

Wen, L. and Li, W. (2013) 打造竞争优势,推动中国汽车产业升级 (Strengthen competitive power and promote China Automobile Industry upgrade), *Auto Review*, **11**, 98–101.

Liu, J. (1999) *One step Aheard: Case of Santana Localization*, Shanghai: Shanghai Finance University Press.

McComb, F. W. (1974) *Veteran Cars*, London: Hamlyn.

Ministry of Finance (2009) 节能与新能源汽车示范推广财政补助资金管理暂行办法 *(Interim Measures on Finance Subsidy for Energy-saving and New energy vehicle Demonstration and Extension)*, http://www.ysxny.com/chanpinjishu/20141203/97200.html. Accessed 27 March 2015.

Ministry of Finance (2010) 关于开展私人购买新能源汽车补贴试点的通知 *(A notice on Finance Subsidy for Private Purchase of Energy saving and New energy vehicles at pilot cities)*, http://jjs.mof.gov.cn/zhengwuxinxi/zhengcefagui/201005/t20100531_320528.html. Accessed 27 March 2015.

Ministry of Finance (2013). 关于继续开展新能源汽车推广应用工作的通知 (A notice on continuous improvement of Energy-saving and New energy vehicle promotion work) [Online]. Available at http://www.gov.cn/zwgk/2013-09/17/content_2490108.htm. Accessed 13 April 2015.

National Bureau of Statistics of China (1949–1984, 2012), *China Industry Economy Statistical Yearbook*, Beijing: China Statistics Press.

National Development and Reform Commission (1994) 汽车工业产业政策 *(Automobile Industry Policy)*, http://www.china.com.cn/zhuanti2005/txt/2004-06/10/content_5583597.htm. Accessed 27 March 2015.

National Development and Reform Commission (2004) 汽车产业发展政策 *(Automobile Industry Development Policy)*, (2nd Edition). Beijing: National Development and Reform Commission. http://www.sdpc.gov.cn/fzgggz/gyfz/zcfg/200507/t20050707_27861.html. Accessed 27 March 2015.

National Development and Reform Commission and Ministry of Commerce (2011) 外商投资产业指导目录 *(Catalogue for the Guidance of Foreign Investment Industries)*, http://www.gov.cn/gongbao/content/2012/content_2144287.htm. Accessed 27 March 2015.

New Scientist (2014) China rules on coal, *New Scientist*, **223**(2987), 20 September, 7.

Nieuwenhuis, P. (2014) *Sustainable Automobility: Understanding the Car as a Natural System*, Cheltenham: Edward Elgar.

Nieuwenhuis, P. and Zapata, C. (2005) Can China reduce CO2 emissions from cars?, *Greener Management International*, **50**(Summer 2005), 65–74.

Shanghai Government (2014) 上海新能源汽车补贴新政策 *2014 (Shanghai Finance Subsidy Policy of New energy vehicles policy 2014)*, http://www.e-bus.com.cn/news/news_210.html. Accessed 27 March 2015.

The State Council (2009) 汽车产业调整和振兴计划 *(Automobile Industry Restructuring and Revitalization Plan)*, http://www.gov.cn/zwgk/2009-03/20/content_1264324.htm. Accessed 27 March 2015.

The State Council (2012) 节能与新能源汽车发展规划*(2012–2020) (Energy-saving and New Energy Automobile Industry Development Plan (2012–2020))*, http://www.gov.cn/zwgk/2012-07/09/content_2179032.htm. Accessed 27 March 2015.

Wang, J. Liu, Y. and Ari, K. (2012) Policies and Effects of 'Ten cities and one thousand new energy vehicles' Project, *Scientific Decision-Making*, **12**, 1–14.

Wen, Q. (2014) 大力兴建充电设施,破解 "最后一公里难题" (Vigorously construct charging facilities to overcome 'the dilemma of the last mile'), *Auto & Safety*, **7**, 74–78.

Xiao, L. (2014) 多网融合疏缓充电桩 "里程焦虑" (Multi-network integration relieves 'driving-range anxiety' between charging points), *China Plant Engineering*, **9**, 18–22.

Xie, W. and Wu, G. (1997) 中国汽车工业的技术引进 (The technology introduction policy in China automobile industry), *Beijing Automotive Engineering*, **2**, 1–9.

Xu, L. (2001) 国家863计划电动汽车重大专项正式启动 (The initiation of 863 Program-----electric vehicle project), *Science & Technology Industry of China*, **3**, 49–50.

Ye, S. (2010) 骨干企业联合行动推动节能和新能源汽车产业化发展 (Cooperation of Top 10 enterprises to promote energy-saving and new energy vehicle development), *Environmental Protection*, **18**, 21–23.

Zhang, P. (1995) 技术优势与跨国公司的产业控制:北京吉普案例的分析 (Technology advantages and multinational corporation: case of Beijing Jeep), *Economic Research Journal*, **11**, 30–39.

Zhengzheng, G. (2006) Green light given to eco-friendly vehicles, *China Daily*, 5 January, http://www.chinadaily.com.cn/english/doc/2006-01/05/content_509279.htm. Accessed 27 March 2015.

Zhang, Y. (2009) 中海油将投资建动力电池生产线 (The investment of CNOOC in building power battery production line), *Beijing Times*, 3 December, B42.

Zhao, C. and Ao, B. (2013) 十城千辆"落空,新能源汽车新征途 ('10 cities and 1000 units' project failed, and an alternative development path is needed), *21st Century Business Herald*, 22 January, 22.

Zheng, Z. (2010) 电动汽车的政策环境及发展进程— "863计划节能与新能源汽车重大项目" (Policy and Development Progress of New Energy – 'Energy-saving and new energy vehicle Project in 863 Programme'), http://wenku.baidu.com/link?url=1LHMx2iosJNCOEekNDuNz1Yz5W7kkbR97RspYoHqrpygCMY3CXkaLUAxLiBWC9oOvWVAe3aNw_sxVOnLxgvQU01ftBCeSS4_9i9QD5XREda. Accessed 27 March 2015.

11

Forging Ahead or Stagnating?: An Analysis of Indian Automotive Industry

Maneesh Kumar
Centre for Automotive Industry Research, Cardiff Business School, Cardiff University, Cardiff, Wales, UK

11.1 Introduction

The Indian Automobile Industry has come a long way since the country's independence and has emerged as a 'Sunrise Sector'. After delicensing of this sector in 1991 (Shastry and Pradhan, 2013), it has transformed itself from being a protected locally focussed industry to one of the fastest growing automotive markets in the world (Arthapan, 2011). The industry is witnessing a phase of rapid transformation and growth due to stable economic growth and infrastructure development (Wipro Infotech, 2013). The Indian government introduced a 10 years Automotive Mission Plan 2006–2016, giving strong support to the automotive industry to increase production, stimulate domestic demand and promote the export of vehicles. The growth of India and China as a global automotive hub can be realized from the fact that Asia's contribution to global production capacity in the automobile industry increased to over 50% in 2012 from 15 to 20% in 2002 (Goyal, 2013). The global automobile production has now shifted from advanced countries to emerging countries like India and China.

China has become the top country in car sales, beating the United States. Foreign players, lured by this huge growth potential, are shifting production bases to Asia to be closer to their customers. India with a capacity to produce three million cars is the third largest (Goyal, 2013). However, this growth in the Indian automotive industry started since the start of liberalization and opening of the Indian automobile industry to foreign players in 1991. The delicensing of Indian Automobile Industry and concurrent induction of foreign players have changed the

The Global Automotive Industry, First Edition. Edited by Paul Nieuwenhuis and Peter Wells.

market dynamics, allowing the majority of the leading automotive manufacturers in the world to establish their presence in the Indian market. The foreign direct investment (FDI) in this sector from April 2000 to June 2012 is over US$ 6.96 billion. The industry provides direct employment to over 500 000 people and indirect employment to over 1 000 000 people in India, contributing 6% to Indian GDP and 21% to India's indirect tax revenue. The liberalization policies and open competition also had a significant effect on the local players in terms of production, marketing, export, technology tie-up, product upgrading and profitability. In the world, in terms of ranking, India occupies 1st position in the three-wheelers category and 2nd position in the two-wheelers category, both of which are classified as automotive in India. It also occupies 5th position in the commercial vehicles category and 7th position in the passenger vehicles segment. The country is also a hub for auto component manufacturing which includes more than 500 firms in the organized sector and approximately 31 000 enterprises in the unorganized sector to cater the needs of vehicle manufacturers (original equipment manufacturers (OEMs)), state transport, defence, railway and export to OEMs abroad and aftermarkets worldwide. To strengthen the position of Indian automobile industry by 2016, the central government in 2006 announced plans to upgrade the social infrastructure including areas, such as highways, railways, ports and the provision of energy, which were seen to be the most serious bottlenecks of Indian economic development (Arthapan, 2011; Sasuga, 2011). The commodity tax on the compact segment was reduced from 24% before 2005 to 8% in 2009.

The next section will briefly discuss the history of Indian automobile industry followed by statistics on the performance of the automobile industry. The author will also discuss the critical factors responsible for the growth of this sector, key challenges faced by the industry and some suggestions on how to tackle those challenges.

11.2 History of the Indian Automotive Industry

This section will give a brief introduction to the history of the automobile industry which has evolved significantly in the last 30 years (Figure 11.1). The growth of the Indian automobile industry can be discussed in three phases – pre-1983, between 1983 and 1991 and post-1991 (Kumaraswamy et al., 2012; Saranga, 2009; Shastry and Pradhan, 2013). There were no local automotive players till the 1940s when Ford and General Motors used to import vehicles to India for domestic use; see Figure 11.2. As identified from the literature, the first car to run on Indian roads dates back to 1897. The early 1940s witnessed the start of a few local companies including Tata Motors (TM), Mahindra & Mahindra (M&M), Hindustan Motors (HM) and Premier Auto (PA), which basically imported parts from General Motors and Ford.

The Industrial Licensing Act (No. 65 of 1951) placed significant restrictions on the expansion of private businesses or new start-ups. Any investment by a private firm, national or international, had to go through a series of interactions with government regulators and politicians. In 1953, another regulation was imposed by the government that allowed the operation of those firms who have manufacturing programs in place – mere assemblers of imported vehicles were asked to terminate their operations within 3 years. Only seven key players managed to get approval from the government to have a manufacturing program in place, including HM, PA, TM, M&M, Ashok Leyland Limited, TELCO

History of passenger vehicles in India

1900	1920	1940	1950–1960	1970	1980	1990	2000
Ford India	Ford India	Hindustan Motors (HM)	HM	HM	HM	HM	HM
	General Motors (GM) India	Premier Automobiles (PA)	PA	PA	PA	PA	PA
		Tata Motors(TM)	TM	TM	TM	TM	TM
		M&M	M&M	M&M	M&M	M&M	M&M
		GM India	Standard	Standard	Standard	Standard	Maruti Suzuki India
		Standard		Sipani	Sipani	Sipani	Volkswagen-Audi
		Fiat India			Maruti Suzuki India	Maruti Suzuki India	Skoda Auto India
						Rover	Mercedes–Benz. India
						Mercedes–Benz. India	Ford India
						Ford India	Honda Siel Cars India
						Honda Siel Cars India	Hyundai Motors India
						Hyundai Motors India	Toyota Kirloskar Motors
						Toyota Kirloskar Motors	Fiat India
						Fiat India	BMW India
						Mitsubishi	Mitsubishi
						GM India	General Motors India
							Volkswagen India
							VE-CVs Eicher

Figure 11.1 History of passenger vehicle industry in India. Source: Compiled from various sources

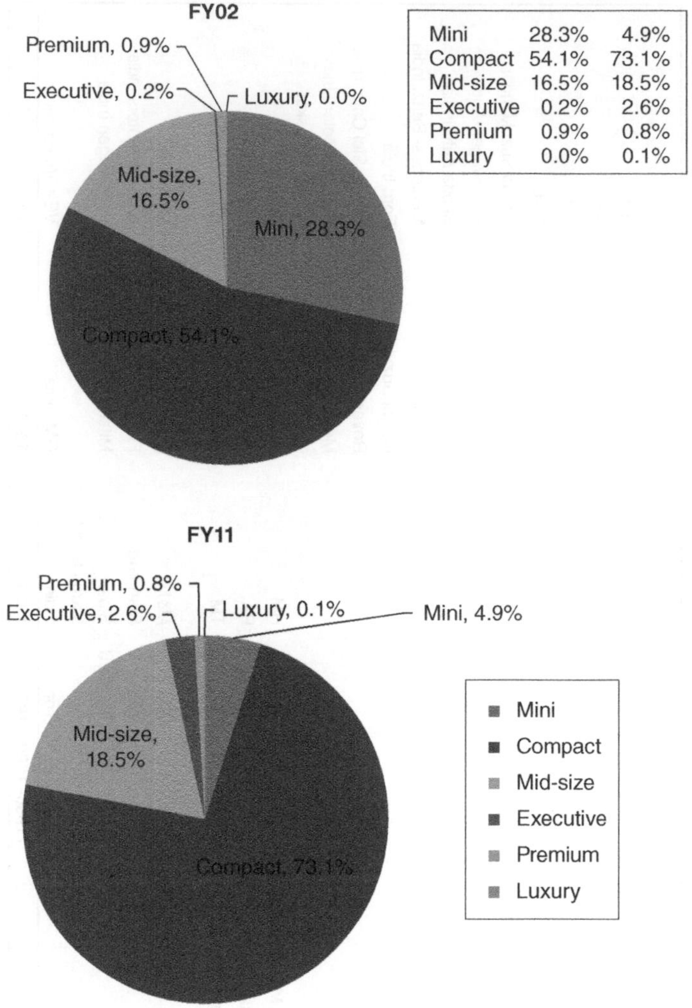

Figure 11.2 Change in car sales composition. Source: Compiled from various sources

(Tata's truck division) and Standard Motors Products of India Limited. Ashok Leyland and TELCO were the key players in the commercial vehicle segment. Not much changed till the end of 1980, except the entry of Sipani Automobiles in the small car market segment.

The early 1980s saw a major resurgence in the Indian automobile car market with the relaxation and changes in government policies including Monopoly & Restrictive Trade Practices Act (MRTP) and Foreign Exchange Regulation Act (FERA) that allowed delicensing of some ancillary products, modifications in licensing policy, concessions to the private sector (both Indian and MNCs) and foreign collaborations policy. The year 1983 saw the establishment of Maruti Udyog Ltd. in collaboration with Japanese firm Suzuki. At the same, HM, PA and SM also introduced new car models in the market.

Following economic liberalization and delicensing in 1991, new industrial policies were introduced that focused more on improving trade and foreign investment, and encouraged technology licensing/transfer (Kumaraswamy et al., 2012). Many foreign players including Hyundai, Toyota, Ford, Fiat, Mitsubishi, Suzuki, GM, Mercedes-Benz, Volvo, Daewoo and Peugeot entered the market. Most of these multinational corporations (MNCs) entered the market by having joint venture (JV) with Indian companies, though these MNCs had more than 50% or even close to 90% stake in the JV.

The entrance of automotive MNCs post-liberalization facilitated upgrading of the overall economy through foreign direct investment (FDI), which also forced the local players to make appropriate investment in improving and enhancing their competencies (Kumaraswamy et al., 2012; Saranga, 2009). Government introduced a uniform policy in 1997 for automotive MNCs to establish not only assembly plants but also manufacturing operations in India. The policy also included other conditions such as investing at least $50 million to establish a wholly owned subsidiary and increasing sourcing of materials from local players. The government also introduced emission norms equivalent to Euro 1 and Euro 2 (see Chapter 11) to encourage local firms to develop their technological and operational capability to match the requirement of OEMs and also meet global standards for export. Further changes in government policies were implemented in 2002, which allowed 100% foreign ownership in the auto and auto components market without the need to involve local players and no requirement of minimum investment in setting up the operations (Kumaraswamy et al., 2012).

This development led to the growth of local prestigious brands such as Tata Motors, operating in the commercial vehicle segment only, to enter the car market in 1998 with the introduction of India's first indigenous developed compact passenger car – Indica. The major breakthrough in recent years was the unveiling of its 'Nano' by Tata Motors that attracted worldwide attention and proved that Indian manufacturers can also compete in design and development of a 'frugal' car through innovative choice of components, materials, engine and design. This was the cheapest new car in the world, also known as One Lakh Rupees car (US$ 2000), developed by an Indian automotive manufacturer. Though Nano received a positive response during its launch in 2009 and sold under the banner of 'world's cheapest car', it has received increasing criticism from the market in compromising its safety element resulting in a number of cars bursting into flames (Bajaj, 2010). Its sales have stagnated in the last few years as the increasing brand-conscious Indian population do not want to associate themselves with a stigma of driving a 'poor man' car (NY Daily News, 2012).

The majority of car manufacturers in India are based around three clusters in the north, south and west. The south cluster near Chennai is the biggest with 35% of the revenue share and has presence of leading passenger car manufacturers including Hyundai, Ford, Renault, Nissan and BMW. The western hub near Maharashtra and the northern hub near Haryana and Delhi respectively shares 33 and 32% of market share (SIAM and ACMA website). Sanand is developing as an automobile hub in the western part of India with the establishment of Tata Motors' Nano plant, followed by Ford India Pvt. Ltd and Maruti Suzuki India Ltd.

The majority of foreign players have now established their production plant in India and are working closely with local suppliers to develop a strong base of local Tier 1 and Tier 2 suppliers to minimize the dependence on import of technology or raw materials (Okada, 2004). This will also help to develop local competence thereby reducing the operational cost due to savings from import and currency fluctuation.

11.3 Statistics on Automobile Industry Performance

The Indian automotive industry comprises four market segments: passenger vehicles (PV), commercial vehicles (CV), two-wheelers and three-wheelers; as pointed out earlier, unlike in other major car producing countries, the latter two categories are also regarded as automotive in India and hence counted in statistics. The domestic automobile industry production figure for 2011–2012 is 20 366 432 vehicles of which the share of two-wheelers, PV, three-wheelers and CV were 76.5, 15.1, 4 and 4.4% respectively, as shown in Table 11.1. The cumulative production data for April–March 2012 shows production growth of 13.83% over same period last year. In March 2012, the overall automobile exports registered a growth of 17.81% compared to same month in 2011. The export figure for the industry during April–March 2012 was 2 910 055 vehicles, registering a significant growth of 25.44%. All the four segments registered a double-digit growth during April–March 2012 – PV (14.18%), CV (25.15%), two-wheelers (27.13%) and three-wheelers (34.41%). Car export crossed half a million mark for the first time in a financial year.

The PV segment is dominated by cars (78%) and MUV/SUV (22%). The major shareholder as per 2012 statistics in the PV category are Maruti (38.6%), followed by Hyundai (14.2%), Mahindra & Mahindra (13.1%) and Tata Motors (10.5%). The world's leading PV manufacturer, Toyota, is placed at 5th position in India with a market share of 6.2%. The two-wheelers market is dominated by motorcycles (80%), scooters (14%) and mopeds (6%). It leads the market share and is expected to continue its dominance in the next decade. The prominent players in the two-wheelers category are Hero MotoCorp (35% of market share) and Bajaj (32% of market share) that captures more than 65% of the market. In the CV segment, Tata Motors lead the market with a whopping share of 62% followed by Ashok Leyland (15%). The majority of market share in the three-wheelers category is shared between Bajaj Auto (42% market share) and Piaggio (41% market share).

As the focus of this chapter is more on PV category, the author will present the statistics of PV sales and growth in 2012 and compare it with performance in 2011. The data and statistics were extracted from SIAM website but verified from other reports on automotive sector. It was for the first time in history that passenger car sales crossed the two million sales figure in March 2012, showing a growth of 19.66% over the same month in 2011. Also, sales growth of total PV in March 2012 was 20.59% compared to March 2011. The PV segment grew at 4.66% during April–March 2012 over same period last year. Passenger cars grew by 2.19%, utility vehicles grew by 16.47% and vans by 10.01% during this period.

Table 11.1 Market share in Indian automobile industry

Domestic market share for 2012–2013	
Passenger vehicles	15.1%
Commercial vehicles	4.4%
Three-wheelers	4.0%
Two-wheelers	76.5%

Source: SIAM, Shastry and Pradhan (2013).

In the passenger car segment, there has been distinct shift in the demand patterns in the last 10–12 years. The demand for the compact cars has grown steadily since 2002, taking the market share of mini cars (see Figure 11.2).

It is interesting to observe contrasting statistics in the sales of different segment cars in India and China market. More than 70% of Indian car market is captured by A- and B-segment compact cars compared to more than 50% dominance by C-, D- and E-segment cars in China (Arthapan, 2011). There is also increasing interest in the high-end/luxury cars due to changing customer preference, increase in purchase power and consumers becoming more brand conscious.

11.4 Stagnation of Industry in 2013–2014

The fortunes of the Indian Automotive Industry have plummeted in the last few years – from a market showing strong performance in FY2010 with a growth of 25% to negative dip in sales in most market segments despite record discounts (The Economic Times, 2014b; The Hindu, 2013a, b). The inventory levels at vehicle dealerships continue to remain very high. High discounts being offered to push sales are negatively impacting the bottom line of automakers and dealers. Domestic PV sales continued to decline at 7.5% during FY 2013 across its three sub-segments (passenger cars, utility vehicles and vans). According to SIAM, there are several factors that have resulted in negative growth of PV industry including a challenging macroeconomic environment, high fuel prices, interest rates and inflation and weak income growth. Apart from the aforementioned factors, the overall situation of the economy and low consumer sentiments have greatly affected demand for the PV industry.

The Indian government slashed the excise duty in February 2014 as a counter measure to boost the sales and growth of the industry. However, the impact of this cut on the growth of the industry seems questionable. The excise duty will only be able to drive volume for the passenger car segment where a price cut of 3–5% can allure a potential customer to buy a car (The Economic Times, 2014a). While the excise duty cut in 2009 stimulated demand, it may not necessarily work this time as the lower price of vehicles due to reduced excise duty will only solve partial problems temporarily. The trade analysts perceived this move by the government, as an indirect strategy to stimulate growth, but it may not inspire confidence among consumers as government will garner more tax revenue, corporate tax and dividend distribution, which offset revenue forgone from excise duty (The Economic Times, 2014a). Other factors affecting the growth include interest rate, depreciation allowance, ability to payback and macroenvironment to decide purchasing decision of an automotive vehicle (The Economic Times, 2014a, b).

In 2013, companies also cut production, expanded distribution further into semiurban and rural markets that continue to do well, increased localization and invested in new product development. Another area to focus attention on is the market for the used car in India, which is underdeveloped. There was an initial attempt made in 2013 to branding of the used-car network and offering warranties.

11.5 Factors Critical to the Growth of the Indian Automotive Industry

The Automotive mission plan 2006–2016 developed by the Ministry of Heavy Industries and Public Enterprises in the Government of India had set the following vision – 'To emerge as the destination of choice in the world for design and manufacture of automobiles and auto components with output reaching a level of US$ 145 billion accounting for more than 10% of the GDP and providing additional employment to 25 million people by 2016'. A roadmap was designed in this mission plan to develop the industry, to provide world class facilities for automotive testing and certification and to provide a level field for healthy competition among the manufacturers including local and foreign players.

The stagnation of the Indian Automobile Industry in 2013–2014 may be an inhibiting factor in achieving the vision set in the Automotive Mission Plan document. However, there are many bright spots that still indicate growth of this sector in the next 5 years. There are several reasons that may support the growth of this industry (Devonshire-Ellis, 2013; Goyal, 2013; KPMG, 2010; Shastry and Pradhan, 2013; SIAM Website; Wipro Infotech, 2013) and are as follows:

- Improvement in living standards and increase in purchase capacity of the Indian middle class due to strong economic growth in the last two decades. This has attracted the major global automotive manufacturers to the Indian market.
- Easier access to finance and cheaper credits for new and used vehicles.
- The availability of trained manpower at a competitive cost is another reason for foreign player to enter the Indian market. Analysts estimate that producing cars in India today may be 15–20% cheaper than in the United States.
- Low cost of operations coupled with skills in design, R&D and manufacturing has led India to be one of the most preferred destinations for investment and manufacturing operations.
- The stagnation of auto sector in markets of United States, United Kingdom, EU and Japan has worked as a push factor for shifting of new capacities and flow of capital to the emerging markets of India and China.
- Large domestic market coupled by strong and improving export performance, for example Tata-owned Jaguar and Land Rover are the most desirable cars in the China market. Jaguar and Land Rover sales increased in China by 48% in 2012 at some 77 000 units.
- The increasing competition in automotive companies has not only resulted in multiple choices for Indian consumers at competitive costs, but it has also ensured an improvement in productivity by almost 20% a year in the industry, which is one of the highest in the Indian manufacturing sector.
- India is ranked lowest in the world's top 10 auto markets in the car per capita ratio (expressed in cars per 1000 population). The vehicle penetration in India is at a low 13 per 1000 people (compared to 45 per 1000 for China). The low vehicle penetration and a growing young population may help India to soon overtake Korea as the second-largest car producer in Asia, after China. The low car penetration when combined with rising income, increasing affordability and easy finance will contribute to an increase in automobile demand.
- The PV demand in the rural areas of India is still under-tapped. The demand for PV in the rural area shows a gradual but steady growth, which is a positive sign for Indian automobile industry to double its total sales in future at the back of steady growth in the rural markets. The PV sale in rural market grew from 6% in 1999–2000 to 10.2% in 2009–2010.

11.6 Challenges and Future of Indian Automotive Industry

The J.D. Power Automotive Forecasting survey in 2011 identified three key deficits that are pulling back the Indian automotive industry: *infrastructure deficit (ID), budget deficit (BD) and trade deficit (TD)* (Arthapan, 2011). The infrastructure deficit is the result of slow industrialization including an underdeveloped road network, power supply and sea ports; undersized manufacturing industries; and infrastructure projects burdened with bureaucratic delays and corruption. In the infrastructure category, India is ranked 90th for the quality of its roads, 83rd for the port infrastructure and 110th for electricity supply (of total 139 nations). The government has already taken steps to minimize this deficit by a plan of spending of $1 trillion in 2012–2016 period in an organized manner, doubling the spending of the previous 4 years. The budget deficit is another concern for India as the national debt has reached 80% of GDP in 2010. This is a worrying factor because it becomes more difficult to borrow money thereby raising borrowing cost. India also needs to tackle the issue of its *trade deficit* due to an imbalance between its export and import related activities compared to other Asian countries like China, South Korea and Japan.

The increasing fuel price has not only increased pressure on manufacturers to develop their technology and processes to manufacture more fuel efficient car but also look for vehicles based on alternative fuels. India needs to catch up with the rising demand for greener and hybrid cars in the export market to establish its presence in the global market. India's contribution in supporting the greener revolution can be witnessed from the evidence presented here: India introduced CNG vehicles in 1998 in Delhi and thereafter in a few other cities, especially popular among transport service providers in the CV and PV segment, to reduce the operational cost and also pollution in the city. The Reva quadricycle was the first electric vehicle to run on Indian roads in 2001 followed by the launch of the first dual-fuel (LPG/petrol) passenger car, Wagon-R Duo, by Maruti Suzuki in 2006, and a hybrid car by Honda in 2008 (Honda Civic Hybrid). Toyota and Honda are currently the leaders in the hybrid car market and other companies like GM, Ford and Nissan are also introducing hybrid car models. However, much more needs to be done in India to improve the fuelling infrastructure before CNG vehicles become more mainstream (KPMG, 2010).

The Global Auto Executive survey conducted by KPMG in 2010 clearly echoed the opinion of most industry leaders to develop alternative fuel technologies in future cars including hybrid technologies and battery electric powered vehicles. The lack of technological consensus may not be helping OEMs in the creation of an adequate green infrastructure. Here government support plays a key role, especially in developing a 10–20 year blueprint for the introduction of Greener vehicles in the Indian market. Government needs to work with OEMs to form public–private partnerships in developing smaller engines using multiple technologies based on CNG, biofuel, hybrids, hydrogen, fuel cells, plug-ins or electric vehicles. It is expected in future that more joint ventures or alliances will be formed between existing local and international players driven by the need for access to better technology, manufacturing facilities, services and distribution network. Example of this includes Fiat supplying diesel engine for Suzuki's vehicles and Tata managing the service and distribution facilities of Fiat India. The presence of global players in the Indian market has favourably altered the landscape of the automotive industry with the adoption of international technologies. There is an increasing propensity towards collaboration between auto companies and IT solution providers with a view to better manage operations and meet critical challenges (Wipro Infotech, 2013). This will help India to catch up with the United States and Europe, as it seems to be 5–7 years behind them in automotive technology and performance requirement.

The auto component sector is going to play a key role to support the future growth of this sector. This can be seen as a challenge and opportunity for this sector to develop and catch up with MNCs in terms of technology, R&D and greener components at an economical price. The auto component manufacturers also need to explore alternative recyclable and lightweight materials to bring down the weight of individual components and thereby vehicle weight that is directly correlated to fuel consumption and CO_2 emission. Bodies like SIAM, ACMA and ACT are supporting OEMs, Tier 1 and Tier 2 suppliers to develop their employee capabilities, operational capabilities and technological know-how capability (Economic Times, 2008; Machinist, 2008).

There are several challenges in improving the growth of Indian automotive sector after a period of stagnation or slump in recent years. However, the factors supporting growth such as improved buying power, low cost operation, skilled labour, improved technological capabilities and development of alternative fuel technologies, under tapped rural market and used car market, will help the Indian automotive industry to achieve the vision of the Automotive Mission Plan and improve its global ranking in all segments of automotive vehicles.

References

Arthapan, M., (2011), 1.2 Billion people. Booming market: Will India be the next China? *J.D. Power and Associates Automotive Forecasting*, McGraw-Hill Companies, New York.

Bajaj, V., (2010), Tata's Nano, the Car That Few Want to Buy, *The New York Times*, 9 December 2010, http://www.nytimes.com/2010/12/10/business/global/10tata.html?pagewanted=1&_r=2& [Accessed 12 February 2014].

Devonshire-Ellis, C., (2013), Why India is Winning the Entrepreneurial Battle with China-360 Degree Analysis, *The Asia Briefing*, 22 July 2013, http://www.fairobserver.com/article/why-india-winning-entrepreneurial-battle-china [Accessed 12 February 2014].

Economic Times, (2008), CII's VSME Programme Mentors Small Businesses for Scaling Up, *Economic Times*, 4 August 2008, http://articles.economictimes.indiatimes.com/2008-08-04/news/28420533_1_indian-smes-sme-sector-manufacturing-sector [Accessed 10 January 2014].

Goyal, M., (2013), Five Reasons Why Auto World Is Shifting to Emerging Markets and What It Means for India, *The Economic Times*, 19 May 2013, http://articles.economictimes.indiatimes.com/2013-05-19/news/39354999_1_auto-mncs-india-auto-industry-automobile-industry [Accessed 12 February 2014].

KPMG, (2010), *The India Automotive Industry-Evolving Dynamics*, KPMG India, Member of KPMG International Cooperative, Switzerland.

Kumaraswamy, A., Mudambi, R., Saranga, H., and Tripathy, A., (2012), Catch-up strategies in the Indian auto components industry: Domestic firms's responses to market liberalization, *Journal of International Business Studies*, **43**, 4, 368–395.

NY Daily News, (2012), Cheap Proves Costly for Tata Nano in Status-Conscious India, *NYDailyNews.com*, 30 January 2012, http://www.nydailynews.com/autos/cheap-proves-costly-tata-nano-status-conscious-india-article-1.1014012 [Accessed 12 February 2014].

Okada, A., (2004), Skills development and interim learning linkages under globalization: Lessons from the Indian automobile industry, *World Development*, **32**, 7, 1265–1288.

Saranga, H., (2009), The Indian auto component industry – Estimation of operational efficiency and its determinant using DEA, *European Journal of Operational Research*, **196**, 2, 707–718.

Sasuga, K., (2011), The Impact of the Rise of Chinese and Indian Automobile Industries, In *The Scale of Globalization. Think Globally, Act Locally, Change Individually in the 21st Century*, 286–291. Ostrava: University of Ostrava, 2011. http://conference.osu.eu/globalization/publ2011/286-291_Sasuga.pdf [Accessed 30 March 2015].

Shastry, T. L., and Pradhan, J., (2013), Indian foreign trade with reference to automobile industry-an analysis, *International Journal of Business and Management Invention*, **2**, 9, 62–71.

The Economic Times, (2014a), Interim Budget 2014: Excise Duty Cuts Across Auto Segment Roll in, But Investors Still Wary, *The Economic Times*, 18 February 2014, http://economictimes.indiatimes.com/articleshow/30580465.cms?utm_source=contentofinterest&utm_medium=text&utm_campaign=cppst [Accessed 2 March 2014].

The Economic Times, (2014b), Vote on Account 2014: Fortunes of Auto Industry have Nosedived in Past Few Years, *The Economic Times*, 10 February 2014, http://articles.economictimes.indiatimes.com/2014-02-10/news/47200639_1_rural-sales-sales-declines-luxury-car-market [Accessed 2 March 2014].

The Hindu, (2013a), Domestic Car Sales dip 8 pc, SIAM Pins Hope on New Govt, *The Hindu*, 10 December 2013, http://www.thehindu.com/business/Industry/domestic-car-sales-dip-8-pc-siam-pins-hope-on-new-govt/article5443836.ece?ref=relatedNews [Accessed 12 February 2014].

The Hindu, (2013b), Auto Sales Slump but New Models Perk up Spirits, *The Hindu*, 26 December 2013, http://www.thehindu.com/business/Industry/auto-sales-slump-but-new-models-perk-up-spirits/article5504898.ece?ref=relatedNews [Accessed 12 February 2014].

The Machinist, (2008), CII Launches Pilot Phase of Visionary SME Programme, *The Machinist*, 7 October 2008, http://machinist.in/index.php?option=com_content&task=view&id=1690&Itemid=2 [Accessed 10 January 2014].

Wipro Infotech, (2013), *Step on the Gas: Steer into the Future of the Automotive Industry in India*, Future Thought of Business, 2013, A Wipro Thought Leadership Initiative, 5th edition, January 2013. Wipro, Bengaluru, India.

12

From Factory to End-User: An Overview of Automotive Distribution and the Challenges of Disruptive Change

Ben Waller
International Car Distribution Programme (ICDP), Cardiff, Wales, UK

'Remember, inbound is king!' – this statement, made by an executive in a research interview undertaken by the author not long after switching from mainstream retail into automotive distribution, helps the newcomer to the sector understand many of the issues faced in the car industry. Like many other manufacturing sectors, carmakers require a distribution channel to get the product to international markets and to provide support for customers who have already bought the product. However, whilst the distribution sector accounts for 30% of the automotive value chain, the sector often attracts far less than 30% of attention, as unlike many other retail sectors, the automotive industry is manufacturing and manufacturer led. This sector is typified by an often intense focus on manufacturing efficiency and product design within a few dominant global manufacturing groups, but by comparison also suffers from a far lesser focus on a relatively unconsolidated retailing and distribution sector. This chapter explains the complex and mature automotive distribution market in which carmakers compete with other industry competitors in sales and leasing, finance and aftersales; the main focus of the chapter is physical vehicle distribution and the structure of the supply chain, but like many other retail and service sectors, automotive distribution faces pressures to respond to changing customer behaviour for online channels, improved customer services through better use of customer and vehicle information.

The Global Automotive Industry, First Edition. Edited by Paul Nieuwenhuis and Peter Wells.

12.1 Shipping and Stocking Cars

Henry Ford once said, 'Any customer can have a car painted any colour that he wants so long as it is black'. That statement is mostly no longer true. Cars are produced in a wide variety per model and this creates the first major challenge for automotive distribution – how do you sell variety of an occasionally purchased product?

How cars are sold has a significant bearing on how cars are shipped and stocked; the longer the delivery chain, the less variety can be offered to the customer and more is sold from the stock. There are significant differences in the product variety offered by brands, partially aligned to their localized production capability and lead times. However, it is also a question of business strategy. Asian brands building in Europe generally offer very low variety per model, and the European premium brands typically offer very high levels of potential combinations, the overwhelming proportion of course will never be built (ICDP, 2012a). A noticeable recent trend has been the increase in variety offering in A and B segment cars (small cars), reflecting the shift in the market since the beginning of the financial crisis in 2008, and this has meant an increase in sales from the order pipeline for the smaller cars, so that more variety can be delivered with correspondingly better margins.

Mass production model means that for many markets, manufacturing takes place some distance from where cars are sold (see also Chapters 2 and 4). There are a number of factors that dictate the degree of localization of final assembly: size of the local market, differences in labour and exchange rates, tariffs on imports and exports and maturity of the market. Russia is a prime example of a market that is in the process of slowly opening up to imports whilst at the same time encouraging inward investment to regenerate and grow domestic capacity. The regime of import tariffs and utilization fees in place, since Russia joined the World Trade Organisation in August 2012, is clearly designed to support local assembly. However, the level of local assembly also has the potential to transform how cars are offered, sold and delivered. Research interviews conducted in Moscow and St Petersburg in 2012 by the International Car Distribution Programme (ICDP) found that dealers considered that their customers are impatient, so expect to be able to buy from stock, and at the same time, expect their cars to come with all features and equipment. A study by ICDP in 2012 suggests that around 35% of customers in the Moscow and St Petersburg areas will wait more than 30 days for a new car from order to delivery (ICDP, 2012a). Compare this situation in Russia with that in Germany, a mature market with a high level of sales from the order pipeline and high levels of domestic production and product variety, where many customers have long expected to wait for the car that they want. The long-term trends in customer willingness to wait for order to delivery for European volume brands (weighted by market share, Citroen, Fiat, Ford, Opel-Vauxhall, Peugeot, Renault, SEAT, Skoda and Volkswagen) are shown in Table 12.1. This willingness to wait in Germany, and a culture of choosing from a wide specification, results in sales from the order pipeline of 62.7% for volume brands in Germany (Waller and Arnould, 2013). This proportion has remained relatively unchanged since the first survey in 1994, where 68% of customers were willing to wait, although there are significant differences between the brands that make up this average, and significant differences between different models. The latter is also influenced by the point in the product lifecycle.

Differences in offered variety, culture and localization of supply therefore have a significant impact on lead times and organization of the finished vehicle supply chain. When a customer is sold a vehicle, dealers can obtain it from several different parts of the supply and distribution system. The percentage splits of total sales in 2011, obtained from

Table 12.1 Percentage of volume brand customers willing to wait over 30 days from order to delivery of their new car (ICDP, 2013e)

	1994	1997	1999	2002	2005	2008	2012
France	65	49	48	41	68	46	48
Germany	71	63	54	51	73	43	65
Italy	36	38	32	29	55	44	29
United Kingdom	19	17	32	37	48	59	35

Table 12.2 Supply chain source of customer sales, by brand type, 2011 for the main four EU markets (ICDP, 2012a)

	Dealer (%)	Transfer (%)	Compound (%)	Order pipeline (%)
Volume brands				
France	37.6	8.1	7.9	46.4
Germany	22.2	5.7	9.3	62.7
Italy	46.5	3.2	8.8	41.5
United Kingdom	25.7	7.3	33.9	33.1
Asian brands				
France	39.2	3.1	32.2	25.5
Germany	45.6	3.1	10.1	41.2
Italy	32.8	3.5	11.1	52.6
United Kingdom	23.9	4.1	36.0	36.0
Premium brands				
France	23.0	8.6	2.8	65.6
Germany	31.6	6.2	4.2	58.1
Italy	32.7	2.2	5.6	59.5
United Kingdom	24.2	15.1	11.1	49.6

different parts of the supply chain, are outlined in Table 12.2. This data represents an annual average based on data from ICDP dealer surveys, NSC and OEM data for 21 brands, weighted by brand and market share. The four categories of supply points are as follows:

- Stock held at the dealership
- Stock physically transferred for a customer from another dealership, using a finished vehicle transporter
- Stock sold when held at an official and manufacturer approved distribution centre or compound
- Order pipeline at the factory or in transit to the market (both new customer orders and stock orders that have been adjusted and converted into sold customer orders).

In some markets, sales are almost entirely made from dealer inventory. In the United States, franchise laws designed to protect allocation are interpreted in a way that entrenches behaviour on where and how orders and stock are managed, promoting dealer stocking, and so many cars are sold directly from the dealer 'lot' and subsequently the level of sales from the order pipeline is estimated at around only 10%.

The level of sales from the order pipeline and different sources of stock also determine the setting of appropriate stock levels. In many markets, 60 days of sales is considered appropriate (e.g. if you sell 1000 cars a year that equates to 167 cars). In the United States, reported dealer stock has fallen from a peak of 74 days in 2008 to around 55 days in recent years (ICDP, 2012a). In main European markets, the total stock has averaged around 55 days for most of the last 20 years, with around 40 days held at the dealer (Ibid). During the crisis, stocks actually fell in Europe as scrappage incentive schemes incentivised demand, although the result was a later spike of inventory in 2011. The Chinese market has shown many signs of overheating in the last few years and several reports indicate that inventory levels that had been kept low for many years by strong demand rose in 2012 from around 45 days to around 60–75 days (Shiruzu, 2012). However, stock level should be adjusted to take account of differences in sales culture and sales from the order pipeline, and by the levels of inventory within the delivery pipeline, the latter largely dictated by distance.

Ownership and funding of unsold vehicle inventory vary significantly. Appropriation is the point at which responsibility for funding the full vehicle asset value transfers from the manufacturer to the dealer, and even within one brand, stock financing and ownership policies vary significantly by market. There are examples of (manufacturer-owned) National Sales Companies such as Toyota GB, which order and own all unsold inventory, and dealers can only order a limited number and variants of showroom display stock. There are hybrid models, such as Mercedes Benz in Germany, where manufacturer owned outlets order and own some inventory for themselves and sales agents, whilst other sales points do take ownership of stock. The traditional wholesale model has largely been replaced in many European markets by consignment stocking. Carmakers offer consignment stocks with financial support which can involve interest free and interest bearing periods; in some markets, such as Italy, banks traditionally have played a more important role in financing dealer inventory, but even in this case, the manufacturer is usually viewed as the lender of last resort.

Some brands in some markets centralize unsold inventory in official manufacturer sponsored compounds and distribution centres. It is typical for most brands in the United Kingdom, whilst generally only the Asian brands that build in Europe take this approach across other European markets, as shown in Table 12.3. Centralized inventory reduces the costs of swapping available stock whilst opening up choice to customers, and for many brands, much of this inventory is either under consignment terms or is fully owned by the manufacturer.

The proportion of the order to delivery lead time for a factory order that is made up of movement of the vehicle after completion of production (often referred to as 'passed to sales') is described as the delivery lead time and is typically broken down into to-market delivery or final in-market delivery.

Table 12.3 Proportion of vehicle inventory held in an official OEM compound or distribution centre (ICDP, 2012a)

	France (%)	Germany (%)	Italy (%)	United Kingdom (%)
Volume brands	13	9	15	46
Asian brands	49	15	39	58
Premium brands	14	26	13	30

- The to-market transport of finished vehicles is managed by carmakers and their regional production and sales management companies, whilst the operations are usually subcontracted. Inter-continental car transports (or deep sea movements) are usually organized by the production market, whilst regional flows (or short sea movements) are usually organized by the regional manufacturing or sales organizations.
- In-market delivery is the final delivery within a market from port or compound to the dealer or direct to the customer (e.g. some large fleets have their own pre-delivery inspection facilities), usually by road, and typically organized by the National Sales Company (the manufacturer owned local sales and network management function with the market) or Importer (a third party contracted as the responsible party for sales and network management, often a larger Dealer business). When this in-market delivery lead time is only a few days, dealers 'call off' or source unsold or unallocated cars for sale from a central stocking point. Sometimes these compounds of strategic unsold inventory are located to serve several smaller markets.

12.2 Retail and Distribution

Manufacturing cars requires detailed planning, a task that is complicated by high product variety, and a mixture of sales from both the order pipeline and from finished inventories. Overall volumes for markets are planned some time in advance, although market allocation and mix can change. Sales planning and allocation involves a discussion between each market subsidiary (contracted importer or wholly owned sales company) and sales planning at the regional level; next the overall production requirement is discussed by the sales planning team with a regional production planning team. This process repeats in a monthly cycle, with several months of supply planned into the future, generally referred to as a production programme. This process sets overall volumes and mix, although there will be varying degrees of flexibility of revision to this plan. Equally, for most brands building in Europe, individual orders can be changed by dealers within the individual order-to-delivery cycle. The degree of flexibility in the mix differs significantly by manufacturer, although for most brands building in Europe, individual orders can be amended relatively close to the date of build. Although it is commonplace for carmaker manufacturing profits to be recorded when finished cars leave the factory, cars do actually need to be sold and handed over to customers, and the market subsidiaries manage the network of sales outlets. To create a sales network, most subsidiaries appoint sales franchises rather than sell directly to private customers. The structure and scale of these franchise networks vary by market and region, as shown in Table 12.4.

For several years, Western Europe has been characterized by decreases in outlets, consistent with the economic pressures experienced, and reductions in sales networks have outpaced that

Table 12.4 Signs of network consolidation in Western Europe and the United States (ICDP, 2013b; NADA, 2013)

Western Europe	Sales outlets	Sales per outlet	United States	Sales outlets	Sales per outlet
2005	49 792	293	2005	21 640	779
2013	47 554	246	2013	17 635	819

in service networks, reflecting the delay between lower new car sales and an ageing and shrinking parc, as well as changes in the competition legislation ('block exemption' – see 12.3) rules. Emerging Central and Eastern European markets have experienced the largest growth in sales networks, and Eurozone crisis markets have experienced the largest decreases. As in the United States, sales per outlet in Russia (813) are much higher than in Europe, although there are significant differences between EU markets – most notably between the United Kingdom (444) and Germany (240).

Despite the focus on new car sales, essential to carmakers, within retail networks, aftersales are the main source of profit for franchised dealers in mature markets, and used cars play a significant role in Anglo-Saxon markets, as shown in Table 12.5.

Most car buyers experience manufacturers' products as used cars, sometime after the first owner has replaced their car. Once cars are sold, they are relatively easy products to move around! The high residual value is such that cars have a long life, typically having several owners across the lifecycle from production to scrappage, each phase of ownership providing business opportunities in sales and aftersales. For example, in the United Kingdom, a car typically has four owners (SMMT, 2012).

Flows of used cars exist between markets due to differences in demand for new or used cars, and where import and transport costs allow a retained margin, as shown in Table 12.6. Due to right-hand drive limitations, the United Kingdom and the Republic of Ireland form effectively a closed submarket within Europe, whilst Poland is an example of a significant net importer of used cars, where demand for competitively priced used cars far outstretches demand for new cars. The used car to new car sales ratio is also an indicator of the level of development of a used car market and infrastructure. However, in the major European markets, used cars of

Table 12.5 Contribution to dealer profit, 2012, 2011 data for the United Kingdom

	France (%)	Germany (%)	Italy (%)	The Netherlands (%)	Spain (%)	United Kingdom (%)	United States (%)
New cars	43	26	52	35	36	27	31
Used cars	9	6	12	6	4	26	29
Service and parts	48	68	36	59	60	47	40

Source: ICDP and national sources (ICDP, 2012b, 2013a).

Table 12.6 Import and export flows of used cars by selected European market and the ratio of used car sales to new passenger car sales within the market in 2011 (ICDP, 2012b)

	Belgium	France	Germany	Italy	The Netherlands	Poland	Spain	United Kingdom
Import	23%	2%	3%	3%	5%	39%	5%	0%
Export	–27%	–5%	–16%	–15%	–14%	–1%	–14%	–1%
Used: new ratio	1.3	2.8	2.2	1.8	3.6	5.8	2.2	3.3

Source: ICDP estimates and national sources (CCFA, DAT, Unrae, Bovag, Autostat, Ganvam, BCA, SMMT).

Table 12.7 Share of used cars transactions

	France (%)	Germany (%)	Italy (%)	The Netherlands (%)	Russia (%)	Spain (%)	United Kingdom (%)
VM networks	22	35	22	28	4	24	23
Independent networks	27	24	22	30	6	18	37
Private	51	41	56	42	90	58	40

Source: From Argus, DAT, Unrae, Bovag, Ganvam, BCA (ICDP, 2013a).
ICDP consolidated estimates for 2012.

5 years or more represent the vast majority of used car sales. The average age of the circulating car parc will also have an impact on the spread of age of cars sold. For example, the average age of the circulating parc in Russia is nearly 12 years (Davies, 2013), in the United States it is just over 11 years (Polk, 2013) and over 8 years in Europe (Schultes, 2014).

Private (peer-to-peer) used car trading is an important component of most markets, and where there is a significant retail market, the role of manufacturer networks varies, as shown in Table 12.7. Private-to-private sales constitute the majority of transactions in Southern Europe, whilst customers in Germany, the Netherlands and the United Kingdom are more inclined to buy their cars from a professional. Independent traders, unconnected to carmakers, play a significant role in the United Kingdom, the Netherlands and France, whilst Germany is the European market where manufacturer networks have the most significant share of used car sales. Peer-to-peer trading is increasingly facilitated by online marketplaces, such as Autotrader in the United Kingdom and Autoscout in mainland Europe.

Equally, the share of the aftersales market between manufacturer and independent repairers and chains varies by market. Some markets, such as the United Kingdom, have well established and developed independent service and repair sectors, and where this is the case, the manufacturer networks have only one-fifth of service volumes, whilst in other markets, manufacturers enjoy the major share of service work, such as Germany, where 55% of service volume is undertaken by manufacturer networks. These differences in share are also reflected in franchise service retention by age of car – the franchise sector only loses dominance for cars over 8 years old – whilst in the United Kingdom, the franchised networks lose market dominance when the original warranty period expires, which for most brands in the United Kingdom is at 3 years. The high level of aftersales retention in Germany helps to explain why outlets survive on lower new car sales per outlet mentioned previously. However, the buffer provided by aftersales is in long-term decline, as in most mature European markets, both repair and maintenance operations and crash repair volumes are declining, for a number of structural reasons, including extended service intervals, falling annual average distances driven, and increased safety and reliability of new vehicles. And prices will not help, as corrected for inflation, the European car parts market is also decreasing in value terms, which will impact both the manufacturer and independent parts channels. Whilst aftermarket parts sales are being sustained globally, due to the expansion of less mature markets, this presents a challenge to distribution structures in Europe, and as a consequence, the numbers of both service and maintenance outlets, and crash repair centres, have been reducing since 2005 and are expected to decline further.

12.3 Changes to the Dealer Model

The distribution business is split from a manufacturer perspective into retail and fleet (ICDP, 2010). The fleet customers buy professionally and these customers include large- to medium-sized companies, public sector organizations and large daily rental businesses, in addition to the manufacturer's own internal fleet of staff vehicles and demonstrators; these fleet services are often contracted through more direct manufacturer or third party led fleet management channels, whilst the retail channel can be broadly defined as providing for customers buying cars and services with their own funds. The franchised dealer is the standard distribution model for the retail channel, which whilst more visibly designed for sales, is in fact, as we have seen, heavily reliant on aftersales and so tends to translate into a franchise business that includes new cars, service and parts. Some sector-specific regulation arguably reinforces the standard model, but there are in fact quite a few choices open to carmakers.

Competition regulation in the European Union is managed through the General Block Exemption (GBE). Recent changes in this area of regulation confirm the rights of brand-owners (i.e. manufacturers) to choose their distribution networks. For new vehicle sales, manufacturers can decide the place ('open point' in the network), standards (what the franchisee must do to retain the contractual right to sell new cars) and number of outlets (the size of the network that the manufacturer wants within a market). If a potential franchisee applies but there is no open point, then the manufacturer is fully entitled to say, sorry there are no sales franchises available. Each sales franchise can have a nominal territory, although the manufacturer cannot restrict active sales (e.g. online sales leads and sales to out of dealer territory). This is the approach that most manufacturers have taken in Europe, although under GBE, a manufacturer could also have an exclusive system which restricts sales territories. The rules are quite different for aftersales, but for a service and parts franchise, the qualitative selective criteria mean that any potential franchisee who meets the franchise standards is entitled to operate a franchise.

However, some manufacturers do own some of their sales outlets. In the European Union, 3% of all sales outlets, including franchise points and sales agents, and 4.5% of main dealer sales outlets (which have direct sales contracts with the national sales company or importer) are owned by manufacturers (ICDP, 2013b). In the United States, state-based franchise protection laws that were designed to protect the interests of dealers have the effect in most US states of prohibiting manufacturers selling direct, making an agency approach impossible. Electric car manufacturer Tesla has tried to challenge this constraint in several states, and other manufacturers find ways around creating direct outlets for customers to experience the brand. For example, Ford operated pop-up (temporary) outlets in San Francisco in 2012, but avoided dealer protection issues by referring all potential customers to dealers in the Bay Area.

Dealer groups, which are multi-site and often multi-franchise businesses, play a major role in automotive distribution, although retailing is not highly concentrated compared with other sectors such as grocery retail (in the United Kingdom, an extreme example, the four largest grocery retailers enjoy around 75% of the market). The market shares of the top 25 dealer groups in various markets are shown in Table 12.8.

Whilst the United States appears to be the least consolidated market, there are some very large groups in some of the more densely populated states and urban areas. However, by 2010, dealer groups from Russia, China and Brazil accounted for 14 of the top 50 global dealer groups.

Table 12.8 Market share of the top 25 dealer groups (ICDP, 2013d)

	United States (%)	Denmark, Norway, Sweden (%)	Russia (%)	United Kingdom (%)	Poland (%)	France (%)	Italy (%)	Germany (%)	Spain (%)
Share	10	40	35	28	22	19	12	11	12

There are significant issues facing dealer networks which raise questions about the number and role of outlets required to provide an effective and efficient retail and distribution system. Across all markets, surveys show that customers increasingly look online for information and are coming into dealerships armed with more information than ever before. Sales networks will respond to buyer behaviour whilst avoiding alienating more traditional buyers. Service networks will have to restructure to adjust to lower volumes, more internalized work as part of inclusive service plans and an 'always-on' relationship with customers. However, investments in outlets in less mature markets are repeating many of the same mistakes. Just as online banking has challenged the need for high densities of high street retail bank branches, online connectivity and changing customer behaviour may make heavy investment in commercial retail property a liability rather than an asset.

On the whole, the sector is relatively conservative, and with a few exceptions, the same franchise dealership model is replicated globally, which means that there is low innovation in retail formats and little multi-branding, except in sparsely populated areas. At the same time, sector leaders such as CarMax in the United States show the scope for more sophisticated approaches to leveraging data as mainstream retailing has been doing for some years. For example, Tesco (Dunhumby, 2014), who use transaction data enriched with loyalty club information to match individual customer transaction data by product and service to stock turn, margins and sales uplift (promotions impact), margins and sales uplift (promotions impact) data. This data is made valuable to the company via a system called 'The Shop' that is in real time, requires no expert systems knowledge, is accessible to all internal departments, and also to external partners such as first tier suppliers who pay for controlled access to certain information. An executive at one carmaker recently interviewed as part of research into inventory management skills said that:

> In the automotive distribution sector, there is a lack of 'industrialisation' in the data reporting and analysis. We are acting as craftsmen and not as industrialists.

CarMax, a large US dealer group with a strong focus on used cars, made investment in systems and processes for data analysis which amounted to the biggest cost in their first 5 years of trading. They analyse what sells and what does not, understand market prices and demand, actively source the cars that represent the best prospects for a reasonable margin and a rapid stock turn, and continually optimize pricing, applying algorithms that automatically revalue and re-price where necessary every car every night. As the founder of the business stated – 'CarMax can forecast the likelihood of any individual car selling within the next 7 days, store by store. The key is to have data captured at every step in the most efficient way'.

In response to declining aftersales, manufacturers and their networks in mature markets have been looking to increase the sale of finance products that either include or help promote the sale of inclusive servicing, as shown in Table 12.9. Of course, the cost of money has been

Table 12.9 Sales penetration of private new car sales with finance sold at the dealership and monthly service plans in mature European markets (ICDP, 2013d)

	France	Germany	Italy	United Kingdom
Proportion of private new car sales with finance sold at the dealership Sources: FLA, ICDP, CCFA, ASF, GFK/AKA, Assofin	41% (2011)	43% (2012)	51% (2012)	74% (2013)
Percent of new cars sold with service packages, ICDP dealer surveys 2012	8%	21%	8%	26%

at a historic low and the question remains as to how sustainable these rates of private sale financing will be when interest rates begin to rise in Europe and the United States.

However, many manufacturers are now convinced that an increasing proportion of retail businesses will resemble the fleet sector, where motoring services are often managed for the vehicle users, and this shift along with a more data-driven approach to retailing is forcing a rethink of the compartmentalization of car sales and aftersales business, and the roles of dealer groups, outlets and national sales companies.

12.4 The Changing Role of Fleets

Business and government sales account for a significant proportion of new car sales in mature markets, and these parallel distribution channels are changing to respond to changing customer demand for mobility. As shown in Table 12.10, despite the industry focus on the retail channel, in many mature markets, private sales of new cars to individual customers represents only a fraction of the market. Manufacturer internal fleet sales can be considerable, particularly in markets with significant employment in production. Equally, sales to dealer networks make up a sizable volume, via demonstrators, courtesy cars and forms of self-registration that tend to occur when underlying market demand is weak, the latter process often a mechanism by which manufacturers' networks transform excess new car supply into 'zero kilometre' used cars (Waller, 2004). Daily rental fleets, particularly the large internationals such as Avis and Hertz, also act as a destination for oversupply, and the role of rental companies in taking new cars, making them cost-neutral through operations and then making profit on the resale, has made them an attractive merger and acquisition target during periods when carmakers have enjoyed the reserves to move out of their core business. Including rental, true fleet and business sales for many mature markets account for between 30 and 45% of new car sales (and some small business sales will still be accounted for within the retail percentages). Market differences are largely dictated by tax regimes and the relative state of the economy. These fleet and business sales often include finance as part of the bundle, and this business is split between what is termed as 'captive' and 'non-captive' sectors. The captive sector is funded by manufacturer-backed finance, either wholly managed by a manufacturer's subsidiary such as Volkswagen Financial services or via a 'white label' joint venture with a major bank.

The financial crisis changed the landscape of the non-captive sector; for example in the United Kingdom, the number of non-captive providers fell from 17 in 2006 to 9 in 2009.

Table 12.10 New car sales split by private buyers and other types, 2008 estimates ICDP (2010)

	Private (%)	Motability (UK only) (%)	Other fleets, government and small businesses (%)	Daily rental (%)	VM networks (%)	VM internal (%)
France	59.6	0.0	17.7	10.8	10.6	1.3
Germany	40.1	0.0	20.6	10.3	20.9	8.0
Italy	61.0	0.0	24.7	4.2	8.0	2.0
United Kingdom	38.2	8.5	29.4	11.1	5.4	7.4

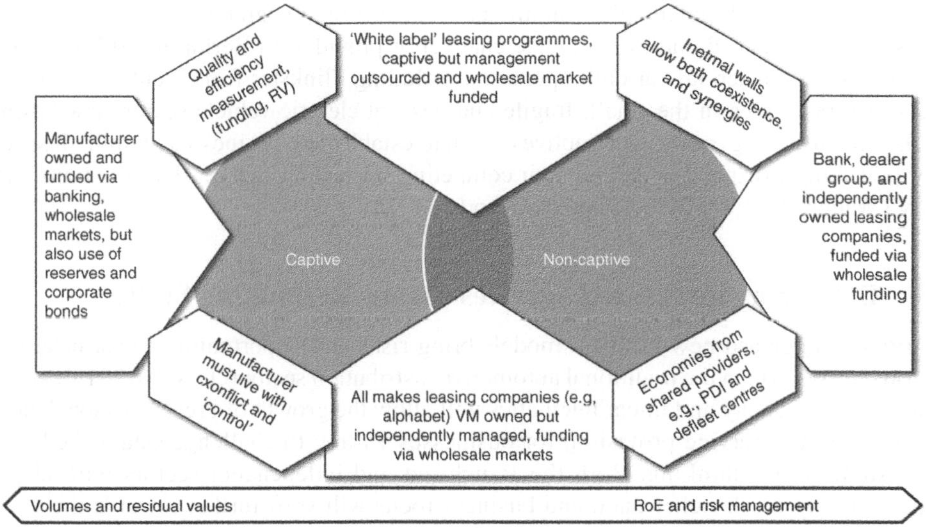

Figure 12.1 The captive and non-captive fleet management and leasing sectors. Source: ICDP (2010)

Manufacturers have clear strategic reasons for financing new car sales that non-captive banks do not (Figure 12.1).

Due to the oversupply element in some fleet segments, as you might expect, residual values per model and brand on average actually decline as volume and fleet share of sales increases, (ICDP, 2010), although some national sales companies manage fleet sales better than others, incorporating residual value management, 'de-fleeting' and remarketing into their approach to new car sales in various fleet segments. There tends to be a link between well-managed fleet operations and sophistication in the management of used cars within franchised networks, as the latter channel is supported with approved used car programmes that help support both residual values for the brand and profitability for the dealer network and fleet sales.

However, the fleet sector itself is going through a period of change, as long standing social models of car sharing enter mainstream services. Within highly populated cities, car use is changing, and this is reflected in the growth of short-term car rental and various forms of car sharing, such as Autolib in Paris, global operator Zipcar that was acquired by Avis in 2013, and carmaker backed mobility models such as Multicity from Citroën, DriveNow from BMW and Sixt, and Car2Go from Daimler Mobility Services. In Europe in 2006, shared car fleet service providers (car clubs) had 200 000 subscribers. By 2011, this had tripled to over 700 000 (ICDP, 2013c). Frost and Sullivan have forecast that this number in Europe will grow to 15 million by 2020 (Ibid) or 1 in every 20 EU driving licence holders, and although all forecasts will be by their nature wrong, it is open to question how wrong – mobile phone take up in the last decade would have surprised many forecasters in the 1980s. Ubiquity of smartphones and other mobile devices is creating new markets for travel and car sharing, a shift that both carmakers and other businesses have both recognized and begun to embrace. As one car club operator noted when interviewed in 2013: 'Integrated apps and transport systems will have a big impact. Ask a young person which they want more, a smartphone or a car – Smartphone, no contest'. The Moovel app from Daimler is an early example of attempts to integrate A to Z travel, potentially a significant prize for the integrator that gains the trust of the customer to be the default first choice channel or provider. In Berlin in 2013, there were ten competing commercial car club operators, including Flinkster from Deutsche Bahn AG. As has also been seen in the small, fragile and nascent electric vehicle sector, new business models are attracting new 'non-captives', often established businesses within their own sectors, wishing to expand or defend their competing interests in market shares in transport or energy (Waller, 2011; see also Chapters 13 and 14).

12.5 Delivering Integrated Services Means Rethinking Skills

Disruptive change and new business models bring risks and opportunities to carmakers and other businesses within the traditional automotive distribution sector (cf. Wells, 2013); always-on mobile connectivity and in-car telematics will allow the growth of greater personalization, and integration of service provision, for businesses willing to challenge established distribution strategy and thinking. Both the franchised and independent sectors will adapt to being driven by much richer data, and business focus will shift further towards the quality of customer interaction and the provision of mobility services rather than just selling cars, parts, maintenance and repair. However, the ability to offer a superior and seamless online and offline customer experience will only be effective if the industry invests in recruitment and training. Furthermore, changing the business cultures within complex organizations, which includes establishing the right people in customer facing roles, is likely to be a greater task and take longer to implement than changing the physical infrastructures and systems (Young, 2012). So whilst traditionally discrete approaches to sales, aftersales and parts have remained remarkably resilient, the legacy of existing business models will come under increasing pressure as the skills and services offered by an automotive distribution business become the key sources of competitive advantage.

References

Davies, M (2013) Russia's red-hot car market stalls. Reuters. 23 December 2013. Available from http://www.reuters.com [Accessed 22 March 2014].

Dunnhumby (2014) Tesco and Dunnhumby case study. Available from http://www.dunnhumby.com [Accessed 22 March 2014].

ICDP (2010) Winning strategies for fleet business. 9 December 2010. Solihull: ICDP. Available from https://www.icdp.net [Accessed 22 March 2014].

ICDP (2012a) Reducing the risks of new vehicle stocking and supply in an uncertain market. Solihull: ICDP. Available from https://www.icdp.net [Accessed 22 March 2014].

ICDP (2012b) From used car dealers to used car retailers: workshop. 05 December 2012. Solihull: ICDP. Available from https://www.icdp.net [Accessed 22 March 2014].

ICDP (2013a) Vehicle manufacturer approved used car schemes: is it time for a refresh? Workshop. 03 September 2013. Solihull: ICDP. Available from https://www.icdp.net [Accessed 22 March 2014].

ICDP (2013b) European car distribution handbook. Solihull: ICDP. Available from https://www.icdp.net [Accessed 22 March 2014].

ICDP (2013c) A Tale of three cities: ICDP Mobility workshop. 17 April 2013. Solihull: ICDP. Available from https://www.icdp.net [Accessed 22 March 2014].

ICDP (2013d) Distribution in a digital age. 20 February 2013 Solihull: ICDP. Available from https://www.icdp.net [Accessed 22 March 2014].

ICDP (2013e) Defining best practice for inventory management in new vehicle supply: ICDP New Vehicle Supply workshop. 19 November 2013 Solihull: ICDP. Available from https://www.icdp.net [Accessed 22 March 2014].

NADA (2013) State of the industry report 2013. Available from http://www.nada.org/nadadata [Accessed 22 March 2014].

POLK (2013) Polk finds average age of light vehicles continues to rise. Polk. 23 December 2013. Available from http://www.polk.com [Accessed 22 March 2014].

Schultes, R (2014) Europe's lower-gear car recovery. Wall Street Journal. 09 March 2014. Available from http://www.online.wsj.com [Accessed 22 March 2014].

Shiruzu, N (2012) Unfamiliar problem in China: unsold cars. *The New York Times*. 25 July 2012. Available from http://www.nytimes.com [Accessed 22 March 2014].

SMMT (2012) Motor Parc survey, 1 June 2012, SMMT, UK Available from http://www.SMMT.co.uk [Accessed 22 March 2014].

Waller, B (2004) Nearly new cars: the solution to a sales shortfall? Solihull: ICDP. Available from https://www.icdp.net [Accessed 22 March 2014].

Waller, B (2011) Developing a new plug-in electric vehicle ecosystem for automotive distribution. 19th International Colloquium of Gerpisa. 8 June 2011. Gerpisa, Paris. Session no: 12.

Waller, B and Arnould, G (2013) Oversupply in Germany has exposed a long term stocking problem. Solihull: ICDP. Available from https://www.icdp.net [Accessed 22 March 2014].

Wells, P (2013) *Business Models for Sustainability*, Cheltenham: Edward Elgar.

Young, S (2012) The future dealer. Solihull: ICDP. Available from https://www.icdp.net [Accessed 22 March 2014].

13

Impacts of Automobility

Peter Wells[1] and Paul Nieuwenhuis[2]
[1]*Centre for Automotive Industry Research, Cardiff Business School, Cardiff University, Cardiff, Wales, UK*
[2]*Centre for Automotive Industry Research and Electric Vehicle Centre of Excellence, Cardiff Business School, Cardiff University, Cardiff, Wales, UK*

13.1 Introduction

The automotive industry can point, with some justification, to the jobs, wealth creation and human mobility benefits arising from the industry and the products it supplies. As such, the industry continues to occupy a pre-eminent position in many national economies and societies, while the car itself is arguably the defining consumer artefact of the era (Nieuwenhuis, 2014). Nonetheless, the external costs of automobility remain diverse and profound. The intention here is not to attempt a sort of cost–benefit analysis of whether or not the automotive industry 'pays its way' in our societies; such an analysis is inevitably largely determined by the nature of the assumptions used to support the calculations. The purpose of this chapter is to provide an overview of the scope and scale of external costs arising from automobility and then to focus in on two important aspects: road traffic deaths and injuries and environmental costs. In both cases, it is concluded that the rate of improvement in new vehicles and other interventions that might be expected to reduce these particular burdens is overwhelmed at a global level by the growth in vehicle ownership and use. In the meantime, of course, the locus of these burdens is also shifting to the newly motorizing nations of the world.

13.2 Externalities and Automobility: A Broad Perspective

Automobility has a long association with key environmental problems, most notably air pollution due to engine exhausts in petrol and diesel vehicles. However, its impacts occur throughout its value chain. First, the industry is a major consumer of virgin raw materials, though over the years the proportion of a vehicle accounted for by different materials has

The Global Automotive Industry, First Edition. Edited by Paul Nieuwenhuis and Peter Wells.

changed substantially and is likely to change further into the future under the pressure to reduce average vehicle weight (EPA, 2012). It also varies widely between vehicles. As Drucker (2012) showed, actual usage of aluminium, as an example, varied in their sample of 57 European vehicles from 75.5 kg for the Fiat Panda to 561.3 kg for the Land Rover Range Rover (with an overall average of 160 kg).

As Taub et al. (2007) show, the material composition of the car has changed substantially during the early years of the industry. In 1906, about 65% of the average new car by weight was wood, yet by 1912 over 75% was composed of low-carbon steel. After the mid-1970s, however, the low-carbon steel content gradually declined as other materials offered lower weight and greater functionality in a range of components and systems. By 2007, low-carbon steel accounted for just over 60% of the average new vehicle by weight. The car has thus become increasingly multi-material, with aluminium, high strength steel, magnesium, rubber, glass and various polymers all expanding their share of total vehicle weight.

While the classic mass production car of the all-steel era featured vehicle bodies and many other components made from steel, it is expected that piecemeal replacement by alternative materials (e.g. plastics, aluminium, organic materials and carbon fibre) will result in multi-material vehicles. Nonetheless, the automotive industry is likely to remain a major consumer of virgin materials because of the qualities understood to be required for production and sale of new cars in high volume. The industry has a long record of demonstrating alternative paradigms, but none has made a significant impact in production. A typical example is the Chrysler Composite Concept Vehicle, shown in 1996, as a 655 kg vehicle that would cost only US$6000 and would have a body constructed from four large injection-moulded thermoplastic sections made of recycled Polyethylene terephalate (PET) bottles (Nieuwenhuis and Wells, 1997, 134–135). In 2013, with vehicles like the BMW i3 coming out into the market, which features what amounts to a carbonfibre body on an aluminium chassis, the prospects for more enduring material shifts are certainly stronger.

Beyond material consumption, the industry generates significant environmental issues in manufacturing (e.g. land use, air and water pollution, noise and energy consumption), albeit much reduced in a per-vehicle sense from earlier periods. In use, vehicles generate more environmental problems ranging from the visual intrusion of endless cars in the urban landscape to concerns over noise pollution, congestion, energy consumption and land use.

13.3 Death and Injuries from Road Traffic

Road traffic deaths have emerged as a global concern as levels of motorization have increased, extending into countries that hitherto exhibited comparatively low vehicle ownership rates per capita (CGRS, 2009; WHO, 2009). OECD countries have, overall, seen a reduction in road traffic deaths and injuries per 100 000 of the population in the years since 2000 (ITF, 2013). Countries that are experiencing a rapid growth in automobility tend to exhibit higher death and injury rates, but countries that have experienced high levels of vehicle ownership and use for many years then tend to show lower death and injury rates (Kopits and Cropper, 2005, 2008). Rapid motorization often proceeds ahead of the capacity of the available infrastructure, both social and physical. The result is a surge in inexperienced drivers, other road users suddenly faced with much greater traffic levels and an inadequate support system in terms of vehicle maintenance and repair, emergency and police services and related matters. All of these factors tend to result in higher road traffic fatalities. Conversely, countries that have experienced automobility for a long time have also adjusted. Traffic types may be more effectively segregated, for example safety legislation is comprehensive and more effectively enforced, and all road users are more aware of the risks.

It is recognized that it is by no means straightforward to determine internationally-consistent road traffic fatality data (Sauerzapf et al., 2010). The World Health Organisation in 2013 reported that an estimated 1.24 million people were killed on roads around the world in 2010 of which 27% were either cyclists or pedestrians (WHO, 2013). Road traffic is the leading cause of death worldwide for people aged 15–29: about 75% of road traffic deaths are of young males. The traditional policy focus has been on vehicle design, infrastructure design, and traffic rules and their enforcement. It is readily apparent, however, that road traffic deaths are not monocausal but arise out of specific combinations of multiple factors that vary across time, place and circumstance. Concerted efforts across all these aspects have at least managed to stabilize the global death rate, while motorization has continued to increase, but as the WHO (2013) report argues much remains to be done.

In terms of vehicle design, mitigation efforts are traditionally divided into passive and active solutions (see also Chapter 11). Passive systems such as crumple zones or seat belts are engineered into the vehicle but do not involve active intervention. Active systems have been subject to much research and development in recent years and can range from items such as airbags and automatic seat belt tensioning systems through to radar-based automatic braking and even alcohol locks that prevent drivers from using the car. As a result, vehicle occupants have a somewhat improved chance of emerging from an incident alive; but of course the physical forces involved can be immense and cannot always be managed by such systems, especially at speeds over about 80 km/h.

In terms of infrastructure design, again progress has been made with, for example, the separation of different modes, the improved design of intersections and junctions, better signage, road calming measures and so forth, all contributing. Moreover, the increased monitoring of highways via remote cameras, the provision of weather warnings and other measures help reduce problems with the daily use of infrastructures.

A major area of progress that is being pursued by the UN and WHO is that of bringing in legislation to control vehicles and drivers with regard to a wide range of issues from the wearing of seat belts to speeding, alcohol consumption and vehicle maintenance. Alongside such legislation, however, is the need for consistent and effective enforcement that can be problematic in many countries. Welki and Zlatoper (2007) show that actual limits cannot be considered in isolation; effective enforcement is needed to ensure a sufficient proportion of road traffic observes the limits. It is well established that road traffic fatalities are closely correlated with higher vehicle speeds, notwithstanding the efforts of vehicle manufacturers to install passive and active safety systems.

Those concerned with understanding road traffic death rate variations have increasingly sought cultural explanations (Factor et al., 2007; Nordfjærn et al., 2012), including some attempts at categorizing countries along sociocultural variables at the institutional level (Gaygisiz, 2009, 2010). In research terms, there has been rather less attention given to the consequences of behaviour and attitude as contributors to the level of road traffic deaths, particularly in so far as variables of that sort are found to be distinctive and enduring features of the ways in which motorization is experienced and expressed in different places (Moeckli and

Lee, 2007; Redshaw, 2008). There appear to be some limits on the effectiveness of the traditional solutions of vehicle designs, infrastructures and legal frameworks, which has prompted the quest for explanations grounded in an understanding of cultural differences as contributors to observed outcomes in terms of road traffic deaths (FTS, 2007; Vereek and Vrolix, 2007). In broad terms, the perspective is that cultural attitudes have a significant contribution to risky road use and driving behaviours (Melinder, 2006; Redshaw, 2008; Molnar et al., 2012), and it is these instances of risky behaviour that contribute strongly to road traffic deaths. One particular area of concern in this respect is that of alcohol consumption, which can lead to risky behaviour by drivers, passengers, cyclists, pedestrians and other road users (Albalate, 2008).

A second area of interest is that of corruption. The approach with respect to corruption is not to focus narrowly on how corrupt officials may contribute to road traffic deaths (Anbarci et al., 2006; ArriveAlive, 2009), although this may be significant, but rather to use a corruption index as a proxy for the tendency of the population of a country to observe legal compliance (Fisman and Miguel, 2006; Wells and Beynon, 2011) and hence whether or not that population is inclined towards risky behaviours.

Finally, the surge of interest in electric bicycles and mopeds is potentially an area of concern. There is already some evidence from China, where such machines have been in more widespread use than in many other countries, that there is a corresponding increase in deaths and injuries (Feng et al., 2009). A key problem is that these e-bikes are too slow and vulnerable for the road but too (comparatively) fast for bicycle lanes or pedestrian paths. While such e-bikes actually offer some interesting potential in terms of sustainable urban and suburban mobility, they do not fit existing regulatory categories and frameworks or cultural norms and expectations. As a consequence, the disruptive influence of such a change to the mobility matrix can have significant consequences in terms of deaths and injuries.

In the longer term, technologies may be introduced – currently experimental – that have the potential to avoid accidents altogether (Mitchell et al., 2010). The 'connected car' will be aware of its surroundings such that it is also capable of avoiding running into those surroundings, including other cars, cyclists, pedestrians and other road users. Such technologies still require further development and may also need additional infrastructures to be created and for those to become more robust than current systems. Until then, making cars safer will continue to be the primary focus in this context.

13.4 Environmental Impacts

Although the dangers of cars were recognized from the beginning, they were initially welcomed as a cleaner alternative to horses. Cars leave little if any solid waste, while the horse system in urban environments created the need to remove horse manure, urine and horse carcasses on a daily basis. By 1900, for example the United Kingdom had some 3.5 million horses in use, of which 1.5 million were in urban areas. As well as forming the crucial first and last mile in logistics systems – railways do not come to each house, shop or factory – the use of private carriages had increased dramatically in the latter half of the nineteenth century on the back of a growing middle class. In addition, horses run on biofuel and the fuel to run each horse could otherwise feed six to eight people. Cars by contrast had none of these problems and, in addition, allowed users to escape the 'miasma' of the city, much as the bicycle had done (Nieuwenhuis, 2014).

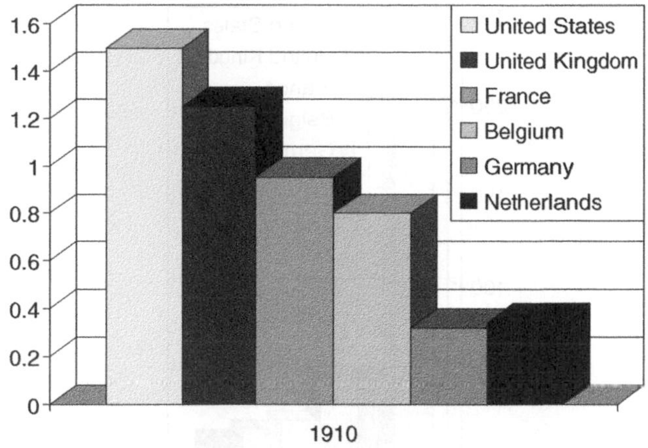

Figure 13.1 Car ownership in key markets 1910 (cars/1000 people). Source: Compiled from various sources

In fact it was felt that the higher speeds allowed the ingestion of greater amounts of clean country air. One could also argue that the increasing awareness of the countryside as a result of car and bicycle use has itself fuelled greater concern for our natural environment.

Nonetheless, some concerns did exist; Hospitalier (quoted in Nieuwenhuis et al., 1992) expressed the belief that internal combustion vehicles had no place in urban environments, instead arguing that electric vehicles were more appropriate. The dust thrown up by cars on unpaved rural roads was also of concern, being also the subject of one of the first environmental measures implemented by the car industry. The 'dustless' Spyker, made in Amsterdam, was highly regarded in the United Kingdom for this reason, prompting *The Standard* to state in its 15 July 1905 issue that 'The new Spyker car is undoubtedly the nearest approach yet attained to perfection in this respect'. Increasingly, paving of roads made this issue less relevant in the industrialized world, although it persists elsewhere, usually well away from urban areas. As car use and ownership increased, the emissions from internal combustion engined cars, which had come to dominate the market since the invention of the 'self starter' in 1912, became increasingly another area of concern. Car ownership in key markets increased dramatically between 1910 and 1938 (Figures 13.1 and 13.2).

13.5 Toxic Emissions

By the 1930s, Southern California had the highest levels of car ownership in the world. This combined with specific natural conditions in the Los Angeles basin to create the first area where the impact of vehicle emissions became evident. As Brilliant (1989, 153) puts it:

> With an irony almost too tragic to be true, the clear blue sunny skies which more than any other factor have been responsible over the years for attracting people to Southern California and for making it a 'motorists paradise', became discolored and poisoned by the waste matter emitted daily by millions of motor vehicles.

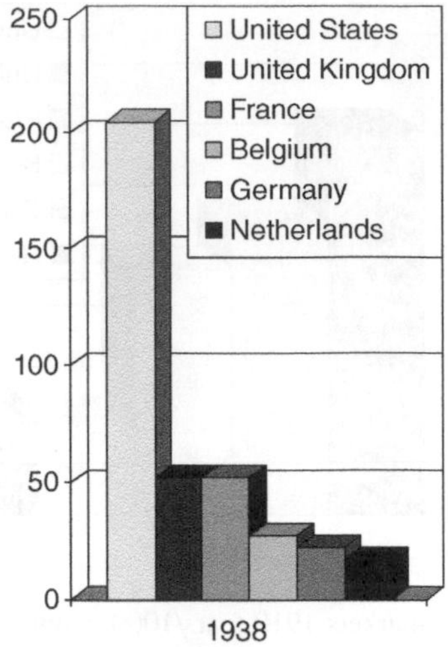

Figure 13.2 Car ownership in key markets 1938 (cars/1000 people). Source: Compiled from various sources

What Southern Californians noticed was due to various toxic emissions from cars, combining both tailpipe emissions with other emissions. With cars of the 1930s, 1940s and 1950s, tailpipe emissions represent about 70% of total emissions, the remainder being made up of evaporative emissions from the length of the fuel system, especially in the hotter climate of this area. Of the exhaust emissions, some 97% were in fact harmless, including water vapour and carbon dioxide, which are not toxins. In addition, the engines emitted oxygen that had not been used in the combustion process and nitrogen that was freed by the heating of air ingested by the engine.

Only 3% is therefore toxic. These toxic emissions consist of carbon monoxide, various oxides of nitrogen, a range of hydrocarbons, particulate matter and sulphur dioxide and trioxide. Carbon monoxide (CO) is a product of incomplete combustion; it is poisonous to humans when inhaled as it binds with oxygen in the blood thereby depriving the blood of oxygen. In the atmosphere, it will turn to carbon dioxide (CO_2) in a matter of hours. The oxides of nitrogen (NOx) generated by the combustion process are produced by the heating of ingested air, rather than any reaction with the fuel used. These consist of nitrogen dioxide (NO_2), nitrous oxide (N_2O) and nitric oxide (NO). Nitric oxide is associated with respiratory problems and acid rain and is also a tropospheric ozone precursor – it reacts to form the photochemical smog identified by Californians. Nitrous oxide is a greenhouse gas 300 times stronger than CO_2. These days, NOx are much more of a problem for diesel than petrol engines and within the EU, where regulation in recent years has focused on reducing CO_2, diesel engines have less severe restrictions on these pollutants than petrol engines. This is not the case in the United States, where toxic emissions and human health have been of primary concern.

Most toxic are the various hydrocarbons which are both unburnt elements of the fuel, as well as resulting from the reaction of the fuel with elements in the air, notably oxygen. However, most of it involves unburnt fuel and can include up to 300 different substances, including aldehydes, such as toluene, xylene and benzene, a carcinogen with no safe limit. These hydrocarbons are primarily a problem with petrol, rather than diesel, which uses different petroleum fractions and is combusted differently. Particulate matter on the other hand is primarily a problem with diesel engine. These have increasingly become associated with various forms of cancer because due to their smaller size, they can be easily ingested and lodged deep inside the lungs. Particulate matter consists of chemicals bound with small dust particles, made up of soot, rubber, unburnt oil and fuel, sulphates and wear debris which are emitted from the engine. They are categorized as PM10 if less than 10 μm in diameter and PM2.5 if less than 2.5 μm. Sulphur dioxide (SO_2) and trioxide (SO_3) are also considered as problematic in diesel engines due to the sulphur content in the fuel; this problem gradually rose from the 1930s onwards (Kitman, 2000). With the more recent advent of low sulphur fuels for cars and trucks, this problem has largely disappeared, although it is still a major problem of marine bunker fuel, something that is only now being addressed. All these have become regulated – as covered in Chapter 11 – however, the first regulation involved crankcase blow-by, a process whereby fuel in the combustion chamber becomes mixed with lubricating oil from the crankcase.

13.6 Current Concerns

As we progress through the second decade of the twenty-first century, two principal threats are of prime importance. First is the question of climate change, the other is the peaking of key resources, notably oil (Heinberg, 2007). Car use represents a major contribution to both these looming crises. The public appears blissfully unaware of the latter, being lulled into a false sense of security by periodic falls in the price of oil; a feature of the price volatility expected after a resource peaks and adjustments are made in the wider economy. Climate change scepticism also appears to be on the rise again, informed perhaps not so much by the science, as by the fact that most people cannot conceive of a world without ready access to their car. A significant part of the answer to both problems is to decarbonize the car and action on this is urgently needed.

We are unlikely to be able to tell unless with very long hindsight whether unusual climate change was happening and we are unlikely to be certain – possibly ever – that human activity was to blame. In fact some argue that our impact on the climate goes back even further (Ruddiman, 2005). In this context, resource constraints potentially have an even more immediate and obvious effect in that rapidly rising oil prices are already having an impact on prices at the pump. Today more than half of all oil consumed worldwide is used to transport people and goods, as well as providing services, making the transport sector as a whole for 95% dependent on oil (Kendall, 2008; Dennis and Urry, 2009). This means that transport and logistics – the lifeblood of our economies – are almost totally dependent on oil. In reality, this constitutes a very high risk strategy that makes the risks taken by our financial institutions in recent years look trivial by comparison. It also makes 'business-as-usual' the highest risk strategy. The current notion is that after oil peaks, the gap will be filled by GTL (gas-to-liquid) and CTL (coal-to-liquid) technologies, thus further perpetuating our dependence on both high carbon energy sources and liquid fuels (Kendall, 2008; Sperling and Gordon, 2009).

Clearly there are other impacts along the automotive value chain, from raw material extraction – for example recovering one tonne of iron ore typically involves the removal of 100 tonnes of lithospheric resources – to the various transport and processing impacts along the manufacturing chain, as well as end-of-life impacts and the impacts from the maintenance and repair of cars. Many of these have been covered elsewhere (e.g. Keoleian et al., 1997, Graedel and Allenby, 1998 and others). Here we have limited ourselves to the more obvious and regulated impacts for practical reasons.

13.7 Role of the Consumer

It is clear that if consumers are not going to give up their cars voluntarily it is more realistic to adapt the car itself. It is also quite clear that the key objective of ensuring the survival of our transport system means reducing its reliance on fossil carbon. We need to change from relying the mineral, lithospheric, sources of the stone-age and embrace the solar age, relying more on renewable energy sources. This decarbonization of the automotive system is the core concern of many regulators. This chapter has analysed the origins of the problems caused by the car, while the next chapter will review the legislative frameworks designed to tackle these.

Traditionally, the car industry has blamed the customer for the nature of the products it makes: 'we only make what the customer wants'. However, the customer is not a car designer or automotive engineer and can only choose from the products offered. Moreover, the customer may be unaware of the implications of his or her decisions; ordinary citizens do not have this information, nor do they have time to track down enough information to make such lifecycle assessments on each and every product they buy or use, although this may change through new ICT technologies (Goleman, 2009). The notion of primary customer responsibility was well and truly challenged by Stuart Hart's influential article (Hart, 1997) where he put the primary responsibility for greening products firmly in the court of the manufacturers:

> Like it or not, the responsibility for ensuring a sustainable world falls largely on the shoulders of the world's enterprises…corporations can and should lead the way, helping to shape public policy and driving change in consumers' behavior.

Hart encourages the industry to help shape public policy not in its own short-term interests but in the longer term social interests that are implied in the sustainability agenda. These will ultimately coincide, of course, a concept that has been recognized by at least some firms. If this partial abdication of customer responsibility seems novel, let us remember that as well as buying today's product offerings, 15 million customers also quite happily bought Ford Model Ts, more than 20 million bought VW Beetles and some even bought BMW Isetta and Messerschmitt Kabinenroller bubble cars. The customer can only choose from what he or she is offered by manufacturers and dealers in the market place. The consumer of automobility does have a role to play. Ultimately we need to recognize that most motorists – aided and abetted by the car industry – are currently engaged in car abuse – an affliction not unlike drug abuse. As with some such activities, moderate use need not be unduly harmful and need to become the norm if we do not want to lose our right to automobility, which in truth is a privilege. We must abandon the automotive excess that has become the mainstream. The issues we need to address go far beyond CO_2 and need to be seen within the broader context of sustainable consumption and production. The car of tomorrow will therefore also need to address the issues of resource depletion, waste generation, congestion and quality of life in the broadest sense. It is likely that the products that will meet such requirements are more involving, more likeable and more fun to drive than the often overspecified, overweight devices of today.

13.8 Conclusions

The automotive industry in the broadest sense, prompted by regulation and public concern, has done much to reduce toxic emissions and is addressing the carbon reduction agenda, although much of the progress here is also prompted by government regulation. There is limited time available to achieve the emissions reductions required, not only to address the climate change agenda but also to deal with the inevitable time when oil is just too expensive to use as a vehicle fuel with current technologies. Radical technologies are available in that they exist in experimental form or are under development, but the inertia in the current mass car production system may not allow such technologies to make the impact needed, wedded this system is to internal combustion technology, liquid fuels, and relatively heavy all-steel bodies. In addition, the link-up between vehicle producers and energy providers is as yet unnatural for both parties, yet electricity stored on-board the car in batteries charged from renewable electricity seems to provide a solution to many of the problems we are facing. Consumers are increasingly willing to embrace lower carbon cars and are also likely to buy plug-in hybrid vehicles and may even take to the pure battery-electric vehicle. However, whether all the stakeholders – regulators, customers, car producers and energy generators – will be able to pull in the same direction remains to be seen. Some of these issues will be explored in later chapters.

References

Albalate, D. (2008) Lowering blood alcohol content levels to save lives: The European experience, *Journal of Policy Analysis and Management*, **27**(1), 20–39.

Anbarci, N.; Escaleras, M. and Register, C. (2006) Traffic fatalities and public sector corruption, *Kyklos*, **59**(3), 327–344.

ArriveAlive (2009) Corruption, traffic enforcement and road safety, Copy obtained from http://wwwarrivealivecoza/pagesaspx?i=2811, accessed 21 May 2009.

Brilliant, A. (1989) *The Great Car Craze; How Southern California Collided with the Automobile in the 1920s*, Santa Barbara: Woodbridge.

CGRS (2009) Make roads safe: a decade of action for road safety, Commission for Global Road Safety, Copy obtained from http://www.fiafoundation.org/our-work/make-roads-safe, accessed 3 November 2009.

Dennis, K. and Urry, J. (2009) *After the Car*, Cambridge: Polity.

Drucker (2012) EEA aluminium penetration in cars, final report (public version), Drucker Worldwide, Copy obtained from http://www.alueurope.eu/wp-content/uploads/2012/04/EAA-Aluminium-Penetration-in-cars_Final-Report-Public-version.pdf, accessed 22 July 2013.

EPA (2012) *Light-Duty Vehicle Mass Reduction and Cost Analysis – Midsize Crossover Utility Vehicle*, Washington, DC: United States Environmental Protection Agency.

Factor, R.; Mahalel, D. and Yair, G. (2007) The social accident: A theoretical model and a research agenda for studying the influence of social and cultural characteristics on motor vehicle accidents, *Accident Analysis and Prevention*, **39**(5), 914–921.

Feng, Z.; Raghuwanshi, R.P.; Xu, Z.; Huang, D.; Zhang, C. and Jin, T. (2009) Electric-bicycle-related injury: A rising traffic injury burden in China, *Injury Prevention*, **16**, 417–419.

Fisman, R. and Miguel, E. (2006) *Cultures of Corruption: Evidence from Diplomatic Parking Tickets*, NBER Working Paper No 12312, Cambridge: National Bureau of Economic Research.

FTS (2007) *Improving Traffic Culture Safety in the United States: The Journey Forward*, Washington, DC: Foundation for Traffic Safety.

Gaygisiz, E. (2009) Economic and cultural correlates of road-traffic accident fatality rates in OECD countries, *Perceptual and Motor Skills*, **109**(2), 531–545.

Gaygisiz, E. (2010) Cultural values and governance quality as correlates of road traffic fatalities: A nation level analysis, *Accident Analysis and Prevention*, **42**(6), 1894–1901.

Goleman, D. (2009) *Ecological Intelligence*, New York: Broadway Books.
Graedel, T. and Allenby, B. (1998) *The Industrial Ecology of the Automobile*, Upper Saddle River: Prentice-Hall.
Hart, S. (1997) Beyond greening: strategies for a sustainable world, *Harvard Business Review*, **75**(1), 66–76, Jan–Feb.
Heinberg, R. (2007) *Peak Everything; Waking up to the Century of Declines*, Gabriola Island: New Society.
ITF (2013) Road Safety Annual Report 2013, Paris: International Transport Forum, OECD.
Kendall, G. (2008) *Plugged In; The End of the Oil Age*, Brussels: WWF.
Keoleian, G; Kar, K.; Manion, M. and Bulkley, J. (1997) *Industrial Ecology of the Automobile; A Life Cycle Perspective*, Warrendale: Society of Automotive Engineers.
Kitman, J. (2000) The secret history of lead, *The Nation*, March 20, Copy obtained from http://past.thenation.com/issue/000320/0320kitman.shtml, accessed 31 March 2015.
Kopits, E. and Cropper, M. (2005) Traffic fatalities and economic growth, *Accident Analysis and Prevention*, **37**(1), 169–178.
Kopits, E. and Cropper, M. (2008) Why have traffic fatalities declined in industrialized countries? *Journal of Transport Economics and Policy*, **42**(1), 129–154.
Melinder, K. (2006) Socio-cultural characteristics of high versus low risk societies regarding road traffic safety, *Safety Science*, **45**, 397–414.
Mitchell, W.; Borroni-Bird, C. and Burns, L. (2010) *Reinventing the Automobile; Personal Urban Mobility for the 21st Century*, Cambridge: The MIT Press.
Moeckli, J. and Lee, J.D. (2007) The making of driving cultures, AAA Foundation for Traffic Safety, Copy obtained from http://wwwaaafoundationorg/home/, accessed 9 March 2010.
Molnar, L.J.; Eby, D.W.; Dasgupta, K.; Yang, Y.; Nair, V.N. and Pollock, S.M. (2012) Explaining state-to-state differences in seat belt use: A multivariate analysis of cultural variables, *Accident Analysis and Prevention*, **47**, 78–86.
Nieuwenhuis, P. (2014) *Sustainable Automobility; Understanding the Car as a Natural System*, Cheltenham: Edward Elgar.
Nieuwenhuis, P. and Wells, P. (1997) *The Death of Motoring? – Car Making and Automobility in the 21st Century*, Chichester: John Wiley & Sons.
Nieuwenhuis, P.; Cope, P. and Armstrong, J. (1992) *The Green Car Guide*, London: Green Print.
Nordfjærn, T.; Jørgensen, S. and Rundmo, T. (2012) Cultural and socio-demographic predictors of car accident involvement in Norway, Ghana, Tanzania and Uganda, *Safety Science*, **50**(9), 1862–1872.
Redshaw, S. (2008) *In the Company of Cars; Driving as a Social and Cultural Practice*, Aldershot: Ashgate.
Ruddiman, W. (2005) *Plows, Plagues & Petroleum; How Humans took Control of Climate*, Princeton: University Press.
Sauerzapf, V.; Jones, A.P. and Haynes, R. (2010) The problems in determining international road mortality, *Accident Analysis and Prevention*, **42**(2), 492–499.
Sperling, D. and Gordon, D. (2009) *Two Billion Cars; Driving Towards Sustainability*, Cambridge: University Press.
Taub, A.I.; Krajewski, P.E.; Luo, A.A. and Owens, J.N. (2007) The evolution of technology for materials processing over the last 50 years: The automotive example, *Journal of Operations Management*, **59**(2), 48–57.
Vereek, L. and Vrolix, K. (2007) The social willingness to comply with the law: The effect of social attitudes on traffic fatalities, *International Review of Law and Economics*, **27**, 385–408.
Welki, A.M. and Zlatoper, T.J. (2007) The impact of highway safety regulation enforcement activities on motor vehicle fatalities, *Transportation Research Part E: Logistics and Transportation Review*, **43**(2), 208–217.
Wells, P. and Beynon, M. (2011) Corruption, automobility cultures and road traffic deaths: The perfect storm in rapidly motorizing countries? *Environment and Planning A*, **43**(10), 2492–2503.
WHO (2009) *Global Status Report on Road Safety: Time for Action*, Geneva: World Health Organization.
WHO (2013) *Global Status Report on Road Safety: Time for Action*, Geneva: World Health Organization.

14

Regulating the Car

Huw Davies[1] and Paul Nieuwenhuis[2]
[1]*Electric Vehicle Centre of Excellence, School of Engineering, Cardiff University, Cardiff, Wales, UK*
[2]*Centre for Automotive Industry Research and Electric Vehicle Centre of Excellence, Cardiff Business School, Cardiff University, Cardiff, Wales, UK*

Chapter 13 highlighted the concerns surrounding the impact of the car in the areas of human health and safety and indeed most of the regulation of cars has focused on these areas of health and safety. However, there are other impacts from the car that have seen less regulation and these include the issue of land use by the car – in cities like Los Angeles and Phoenix, more than half the land area is given over to car-related infrastructure – and social effects, such as new roads separating communities. The issue of deaths and injuries caused by car use is clearly of concern and is outlined further (see also Wells, 2007, 2010 and Chapter 13). Regulation concerning safety has largely focused on the occupants of the car, even though most victims are outside the car. Only recent EU legislation on pedestrian protection has begun to address this through regulation. As Wells (2007) argues, the impact on families and communities from the death or injury of a breadwinner can be devastating, especially in newly motorizing countries, where there is often little redress. The most puzzling aspect of all this is how it became acceptable to kill and injure people with cars right from the early days of motoring. In most countries, killing or injuring a person with a car is seen as an 'accident' endowing it with a degree of inevitability it does not deserve.

14.1 Regulating for Safety

Vehicle safety has long been an area of concern to both consumers and governments. Within Europe, statistics show that cars have become increasingly (crash) safe during the last few decades, driven by a 30 year effort of sustained research, industry innovation, applied legislation and consumer awareness initiatives. The latest passenger cars on the market in the United

The Global Automotive Industry, First Edition. Edited by Paul Nieuwenhuis and Peter Wells.

States, Europe and Japan are safer than ever before, through the development of technologies promoting crash protection (air bags, crumple zones, intelligent seat belts systems, etc.) and new crash avoidance systems (e.g. electronic stability control).

However, globally there is much variation in the provision of vehicle safety equipment from region to region. Currently, vehicles typically get their certification or type approval within the framework of certification procedures established by respective governments on a national level. There are at present no global standards for vehicle safety. Regionally some models may be sold with safety equipment in one country, but with a lower specification in others, if the equipment is not required in legislation.

14.1.1 Development of Vehicle Standards

A working party (WP29) was established in 1952 under the United Nations Economic Commission for Europe (UN/ECE) to draft and implement uniform vehicle standards for Europe. In 1958, the UN/ECE 1958 Agreement was concluded under the auspices of the Economic Commission for Europe with the aim of eliminating distribution barriers that existed in Europe. The purpose of the UN/ECE 1958 agreement was to establish uniform standards on safety for the vehicle's construction and equipment and to implement reciprocal recognition of the equipment certified in accordance with the uniform vehicle standards. The standards on a vehicle's construction and equipment were embodied as UN/ECE regulations in the attachments to the UN/ECE 1958 Agreement.

What made the UN/ECE 1958 Agreement significant was that it allowed signatory countries to reciprocally recognize certification by other member countries, provided that the relevant parts and systems meet the set performance standards. Reciprocal recognition means that certification by a member country eliminates the need to get certification from other member countries and makes it possible to avoid costly and time-consuming duplication of certification procedures with great benefits to internationally traded products such as vehicles.

Whilst the UN/ECE 1958 Agreement was originally created for the UN/ECE region, it was revised in 1995 to form the World Forum for Harmonization of Vehicle Regulations and to expand the scope of its member countries beyond Europe to other UN member countries. Japan was the first non-European country to accede to the UN/ECE 1958 Agreement in November 1998. Australia followed in 2000, South Africa in 2001 and New Zealand in 2002. The World Forum for Harmonization of Vehicle Regulations is the only one of its kind to discuss and develop uniform vehicle standards and regulations concerning the reciprocal recognition of the standards. The signatory nations to the UN/ECE Agreement are empowered by ECE regulations to implement reciprocal recognition. To date, more than 120 UN/ECE regulations have been issued for implementation in the member countries.

14.1.2 European Directives

Within Europe, all new car models have to meet specific safety requirements, referred to as 'type approval', before they are allowed on the road. At present, there are two systems of type approval for high-volume vehicles in Europe. One system is based around UN/ECE regulations and provides for the approval of vehicle systems and separate components but not whole vehicles. European countries can adhere to UN/ECE regulations either voluntarily or mandatorily if a country decides to incorporate the regulation into national regulation. The other system is based around the EU type approval framework Directives. Directives are used to bring different national laws into line with each other and are particularly common in matters affecting the operation of the single market (e.g. product safety standards). Unlike UN/ECE regulations, EU Directives are legally binding on EU member countries. The EU acceded to the UN/ECE 1958 Agreement in 1997 and formally incorporated UN/ECE regulations as EU Directive substitutes.

The most important vehicle safety Directives have been the introduction of crash tests for frontal impact protection and side impact protection to car occupants, as well as, more recently, subsystem tests for pedestrian protection and a Directive on the general safety of motor vehicles.

14.1.2.1 Frontal (96/79/EC) and Side (96/27/EC) Impact Directives

The frontal and side impact Directives have led to rapid developments in car occupant protection in Europe, but the Directives do not specify a particular technology to meet the requirements. The Frontal Impact Directive 96/79/EC for the protection of occupants in head-on collisions applied to new car types from October 1998 and all new cars from October 2003. It tests a car that represents the worst case for the particular model, which is tested using an adult sized (male) dummy. Under the 1958 agreement, UN/ECE regulation 94 provides the same requirements as Directive 96/79/EC. The Side Impact Directive 96/27/EC for the protection of occupants against lateral collision applied to new car types and new car registrations from October 1998. It tests a car that represents the worst case for the particular model using an adult sized (male) dummy. In side impacts, contact with the interior of the car is difficult to prevent, so improvements rely on devices such as side airbags, side impact bars and padding. Under the 1958 agreement, UN/ECE regulation 95 provides the same requirements as Directive 1996/27/EC.

14.1.2.2 General Safety of Motor Vehicles Directive (661/2009/EC)

Directive 661/2009/EC on the general safety of motor vehicles consolidated whole-vehicle type approval requirements. The general safety regulation simplified the previous type approval framework by consolidating 50 separate Directives. The Framework Directive 661/2009/EC also lists a series of UN/ECE regulations that are considered equivalent to or have superseded certain of the separate technical Directives, and proving compliance with these regulations forms an acceptable alternative to compliance with the relevant Directives. It should be noted that where vehicle manufacturers have had the choice of meeting either a European Directive or a UN/ECE regulation, the latter is preferred as UN/ECE regulation allows them access to markets outside the EU. The new type approval Directive also mandates the fitment of electronic stability control (ESC) in all new vehicle types from 2011. Directive 661/2009/EC does not relate to pedestrian protection.

14.1.2.3 Pedestrian Protection Directive (2003/102/EC – Repealed)

Pedestrian protection aims to reduce injury to pedestrians and other vulnerable road users. A draft legislative proposal for a Directive on safer car fronts for pedestrians was prepared in 1992. In 2001, European, Japanese and Korean Automobile Manufacturers Associations committed themselves to a voluntary agreement to improve pedestrian protection. The voluntary agreement on safer car fronts was criticized for implementing weaker test methods that offered a lower level of protection than those proposed by the EU. In 2003, Pedestrian Protection Directive 2003/102/EC set the safety requirements to reduce injury to vulnerable road users if they are hit by the front of a motor vehicle. Pedestrian Protection Directive 2003/102/EC applied mainly to the bonnet and bumper. The Directive introduced pedestrian protection requirements in two stages. Pedestrian Protection Directive 2003/102/EC Stage 1 requirements were required to be met by all new vehicle types from October 2005 and for all new vehicles from December 2012. Manufacturers claimed that it was not possible to reach the Directive 2003/102/EC Stage 2 limits. Directive 2003/102/EC specified that a feasibility study of the stage 2 limits would be conducted using data from testing of the stage 1 requirements. Directive 2003/102/EC review was completed in 2007 and the injury limits were relaxed.

14.1.2.4 Pedestrian Protection Directive (2005/66/EC – Repealed)

In 2005, Directive 2005/66/EC replaced Directive 2003/102/EC. Frontal protection systems Directive 2005/66/EC introduces revised stage 2 limits for pedestrian protection and mandates Brake Assist systems. For Directive 2005/66/EC in the case of new registrations, passenger cars must be fitted with Brake Assist systems by February 2011, meet stage 1 requirements by December 2012 and meet stage 2 requirements by February 2018. New passenger cars undergoing the type approval process must be fitted with Brake Assist systems and comply with stage 1 limits from November 2009 and comply with pedestrian stage 2 limits by February 2013. Pedestrian protection Directive mandates the fitment of ESC systems to passenger cars with effect from November 2011 for new types and November 2014 for all new cars.

14.1.2.5 EU Regulations

Regulations are the most direct form of EU law – as soon as they are passed, they have binding legal force throughout every Member State. They are different from Directives, which are addressed to national authorities, who must then take action to make them part of national law. The replacement of Directives by Regulations simplifies the process of adoption of new and amended proposals by Member States. The proposal is to replace Directives with Regulations. This process is ongoing. In the area of pedestrian safety, Regulation (EC) No. 78/2009 was passed. The main objective of the regulation is to strengthen the requirements aimed at improving the safety of pedestrians and other vulnerable road users in case of injuries resulting from a collision with a motor vehicle. For that purpose, Directive 2003/102/EC relating to the protection of pedestrians and other vulnerable road users before and in the event of a collision with a motor vehicle and Directive 2005/66/EC relating to the use of frontal protection systems on motor vehicles were replaced by this regulation.

14.1.3 US Federal Motor Vehicle Safety Standards

In the United States, the National Highway Traffic Safety Administration (NHTSA) is responsible for motor vehicle safety standards. The United States has a self-certification system.

Self-certification is a system in which vehicle manufacturers undertake to certify that their vehicles conform to federal motor vehicle safety standards and launch them for sale in the market. Under self-certification, the US government carries out a random check about the conformity in the market to ensure the safety of the vehicles. The United States has decided not to accede to UN/ECE 1958 Agreement because of the country's self-certification system on vehicle safety. The United States has not joined the UN/ECE agreement and does not recognize UN/ECE approvals. The United States thinks that it is important to harmonize the vehicle standards internationally and proposed to harmonize the technical standards eliminating, however, reciprocal recognition of certification.

In 1998, the World Forum for Harmonization of Vehicle Regulations adopted the UN Global Technical Regulations Agreement on Motor Vehicles. The 1998 Global Agreement is open to countries that are not part of the UN/ECE 1958 Agreement and seeks to promote international harmonization through the development of Global Technical Regulations (GTRs). GTRs support the harmonization of technical standards eliminating, however, reciprocal recognition of certification. Harmonization of vehicle safety standards is expected to lead to a greater commonality of components, reduced cost of development and production, and simpler certification procedures for the countries involved, all of which will contribute to an increased area of circulation for vehicles and a greater freedom of choice for consumers.

14.2 New Car Assessment Programmes

Regulation has often provided a minimum standard above which a 'market for safety' has developed, helped by new car assessment programmes (NCAP). The first such NCAP was launched in 1978 by the US National Highway Traffic Safety Administration. Consumer information based on crash tests started in Europe in the late 1980s when the UK Consumers' Association and Vehicle Safety Consultants developed a safety rating system for cars. Consumer information provides prospective car buyers with factual information about the safety performance of cars in crashes. Consumer test programmes grade performance as opposed to assigning a simple pass/fail outcome.

The EuroNCAP programme (http://www.euroncap.com/home.aspx) for new cars was introduced by a consortium including the FIA, in 1996, following similar test programmes in the United States (USNCAP). Euro NCAP provides both consumers and manufacturers with independent assessments of the (crash) safety performance of the most common car models in Europe. It was followed by Australia (ANCAP), Japan (JNCAP) and Latin America (LatinNCAP). There are now nine NCAPs or similar bodies active in Asia, Australia, Europe, Latin America and the United States. Some NCAPs include additional pole, whiplash and pedestrian tests, and include crash avoidance systems, such as ESC, in their programmes.

Since 1997, the European NCAP (EuroNCAP) has provided star ratings of the performance of different cars in dynamic tests, which include full-scale frontal and side-impact tests, front-end component tests for pedestrian protection and sled tests for whiplash prevention during rear-end crashes. In 2001, an assessment of seatbelt reminder systems was added. In 2003, a child protection rating was introduced based on restraints for an 18-month old and a 3-year-old child. In 2008, a rear impact (whiplash) protection assessment was incorporated. Since 2009, EuroNCAP points can also be earned for the presence of devices for the prevention of crashes (primary safety), such as ESC and speed limiters and it can be expected that this will result in a faster implementation of these devices.

Until 2009, the EuroNCAP crash safety of children and pedestrians was also tested, but the results were not included in the final assessment. Overall rating from EuroNCAP 1–5 stars was introduced in 2009 to encourage manufacturers to focus on the overall safety of a vehicle, rather than concentrating on achieving good scores in adult occupant protection. To encourage further progress in pedestrian protection from 2012, EuroNCAP requires a minimum 60% score in the pedestrian tests for new cars to receive a 5-star rating. Most vehicles now gain a maximum 5-star rating on EuroNCAP crash tests.

NCAPs help to create a 'market for safety' by raising awareness among the car buying public and rewarding manufacturers that build the safest cars. Over the last 15 years, tests and protocols used by the EuroNCAP in safety ratings, which promote and reward good and best practice, represent the global state-of-the-art in approaches to provide better protection in car crashes. It is an example of market demand driving standards beyond what is required by official regulation; the fact that consumers are familiar with the EuroNCAP test results then encourages the car manufacturers to produce cars that are often safer than is required by law. In 1997, carmakers said 'the EuroNCAP assessment criteria are so severe; no car will ever be able to achieve 4 stars in Adult Occupant Protection'. In 1997, the Volvo S40 achieved EuroNCAP 4 stars. On the other hand, Rover was effectively forced to withdraw the Rover 100 from production due to its poor performance in the EuroNCAP tests. As a consequence, individual manufacturers have introduced different vehicle safety measures in advance of legislation or in response to consumer information programmes.

Monitoring shows that EuroNCAP has contributed to marked improvements in crash protective design to protect vehicle occupants with crash tests, which are generally representative of the types of crash scenarios found on Europe's roads. A study by the European Transport Safety Council suggests that car manufacturers monitor EuroNCAP test results closely and seek to improve poor performing models. The European Commission believes that EuroNCAP has become the single most important mechanism for achieving advances in vehicle safety. EuroNCAP is backed by the European Commission as well as motoring and consumer organizations in every Member State.

14.3 Future Developments

Due to both mass and rigidity, safety for occupants increases with vehicle mass, whereas safety for the occupants of the crash opponent car decreases with vehicle mass. Mass incompatibility is a strong determining factor in the outcome of a crash, and heavier cars are at an advantage in a crash with a lighter car in terms of protection of the occupants. This incompatibility problem is all the more important because of the trend to make cars smaller and more lightweight for environmental reasons. Incompatibility has not been allowed for in the regulation or consumer tests. EuroNCAP crash tests mainly report on passive safety measures and compares vehicles in the same class. Hence, the number of stars gives good insight into the safety within the same model and size class but not between the various classes. Euro NCAP and various other road safety organizations are now discussing how to deal with these differences in mass between cars – an issue that has been explored by some manufacturers.

As long as the problem of incompatibility exists, the increased attention paid to primary safety, and, subsequently, the prevention of crashes is an excellent step. Active safety systems (e.g. driver sensing systems to tell if the driver has been drinking or is falling asleep) continue to grow in importance and in future will be linked to infrastructure management and collision avoidance radars. Brake Assist systems automatically apply maximum braking power when the driver makes an emergency stop. ESC is an extension of antilock brake technology, which has speed sensors and independent braking for each wheel. It improves the safety of a vehicle by detecting and minimizing skids.

Safety devices such as seat belt reminders, ESC and speed limiters are important for ultimate safety because they reduce injury or even help to prevent crashes. Primary safety is a positive development, as preventing crashes is preferable to reducing the severity of the injury in case of a crash. A new 'road map' is being implemented to allow emerging crash avoidance technologies to be included (albeit not supplanting crash protection measures) into the EuroNCAP assessment scheme by 2015. With the rapid deployment on to the market of new technologies, evaluation of systems with reference to real-world crash analysis is essential before wide-scale deployment. Until now, no test methods have been available to determine the actual safe functioning of primary safety, which is a limitation.

14.3.1 Impact of New Vehicle Technologies

Whilst similar to conventional automobiles, electric vehicles have their own particular safety problems. Although electric vehicles have over a 100 years of history, the current status of international energy and environment concerns mean that standards and regulations for electric vehicles are at present hotspots for many countries, institutions and international organizations for standardization (ISO). Both domestic and international organizations have released safety standards for electric vehicles, which have also been added to and complemented along with the development and research of electric vehicles. A large quantity of standards pertinent to electric vehicles have been developed and discussions started in 2010 to develop a new Global Technical Regulation (GTR) concerning the safety of vehicles with hydrogen propulsion.

It is acknowledged that assessing only secondary safety is not enough when it comes to the total effect of vehicle measures for road safety. It is more effective for road safety to prevent crashes than to reduce the severity of injuries of car occupants in crashes. It is, however, important to keep secondary safety at least at the same level. More recent legislation has involved a progressive tightening of existing standards, which are not linked to specific technologies and therefore the cost implications are less clear. However, affordability matters. If the safest vehicles are beyond the reach of the consumer, they will remain in less safe vehicles for longer. It is also important to develop policies that make safer vehicles less expensive, through incentive schemes, global harmonization and fleet purchasing.

Motor vehicles and their components have increased their circulation on the world market as internationally traded products, and it is increasingly necessary to look at the requirements of the society for vehicle safety from a global perspective. There is a new demand to harmonize different national vehicle safety standards internationally. There is a new demand to introduce 'reciprocal recognition', in which certification by one country is reciprocally recognized between exporting and importing countries and regions. Global work will increase the convenience of manufacture and removal of barriers to trade. International standardization through organizations such as UN/ECE, EU, SAE, ISO, IEC and so on, and intergovernmental organizations such as UN/WP29 and standardization organizations of countries and districts which are advanced in automobile industry are all active on this issue.

The monitoring of the performance of European vehicle safety legislation in real crashes to identify progress, as well as future priorities for vehicle safety, has taken place systematically in European countries. The main scientific conferences for international information exchange on vehicle safety policy and research are Enhanced Safety of Vehicles, STAPP car crash conference, International Research Council on the Biomechanics of Injury and Association for the Advancement of Automotive Medicine. Most European countries are represented in technical committees of the UN/ECE and the EU associated with the development of vehicle safety standards and legislation. France, Germany, Spain, Sweden and the United Kingdom contribute to the work of the various working and steering committees of the European Enhanced Vehicle-safety Committee (EEVC). The EEVC brings together national experts and governmental representatives from several countries. Legislative and policy work on vehicles at EU level is led by the European Commission's Directorate of Enterprise and Industry. The Directorate of Mobility and Transport – the lead EC agency for road safety – also plays a key role. Vehicle safety promotion is also pursued by the European Commission through initiatives such as DG Transport's EU road safety action programme and DG Information Society's Intelligent Car initiatives. The European Commission exercises the right to vote in the World Forum (WP29) on behalf of the EU and its 27 Member States. At the same time, the EU retains its ability to legislate independently of UN/ECE where there is a need for earlier or more stringent action.

Market forces tend to produce more rapid responses in individual product design. Manufacturers place safety as a high priority and use it as a means of differentiation, proof of technology leadership and branding. Individual manufacturers have introduced different vehicle safety measures without legislation, in advance of legislation or in response to consumer information programmes, especially in recent years. As part of the European Commission's industrial policy, the CARS 21 (Competitive Automotive Regulatory System for the twenty-first century) process launched in 2005 made recommendations for the short-, medium- and long-term public policy and regulatory framework of the European automotive industry. The European Motor Vehicle Working Group is an advisory group of EC DG Enterprise and Industry, which brings together representatives of the European Commission, Member States and non-governmental and trade associations to discuss proposals for new Directives and standards on vehicle safety.

14.4 Legislating for a Cleaner Environment

Environmental legislation of the car originated in California around 1960, and until the mid 1990s, California remained the leader in this field. Initiatives and regulations started in this state usually formed the basis for federal legislation and subsequently regulation in other countries. The history of vehicle emissions legislation starts more from a concern about human health than about the survival of 'the planet'. The concern about the impact of motorized vehicles on our environment centres around five general issues: emissions, energy consumption, noise, congestion and land use. All of these have been subject to regulatory and legislative control somewhere in the second half of the twentieth century; however, historically most of the debate and regulation has focused on toxic emissions of carbon monoxide

(CO), oxides of nitrogen (NOx), various hydrocarbons (HC) and particulates (PM), with the United States, particularly the State of California and its powerful Air Resources Board, leading the way.

Over time, these US standards influenced standards in other countries. Some adopted them without modification, while others, such as Japan, developed their own standards and their own test cycle. Test cycles, such as the New European Drive Cycle (NEDC), simulate a typical drive and differ in different parts of the world, though they rarely manage to come close to real-world driving conditions, something the EU is hoping to address with a new improved test cycle. Manufacturers thus end up designing their cars to perform well in the test cycle, rather than in real use. Consumers often fail to match either officially declared fuel consumption or emissions performance of the car they bought. Despite such differences, it can be said that worldwide emissions standards are still strongly influenced by the United States, and particularly California. However, in recent years, the lead has shifted more to Europe, as EU standards (e.g. Euro 1–6) are increasingly being adopted by other countries, whose cars are often more similar to European cars compared to North American cars and light trucks. This regulatory trajectory introduced new technologies, such as the catalytic converter and engine management systems, and has led to cars that emit much cleaner exhaust fumes such that the toxic emissions of a modern car are a fraction of what they were from cars before regulation.

By the late 1980s, it was realized that any further improvements in automotive emissions would require the cooperation of both the car and oil industries. For example, catalytic converters are damaged by leaded petrol, so lead had to be removed from petrol before catalytic converters could be introduced. This so-called Auto-Oil approach, involving both these industries and the regulator, was first developed in North America; however, the EU followed in the early 1990s. This became part of a trajectory of gradually tightening limits on toxic emissions by distinct steps. Similar stepped changes in the maximum emissions permitted occurred in Japan, closely linked to the US standards – not least because the United States was and remains a key market for Japanese car exports.

14.4.1 Fuel Economy: Incentives and Disincentives

In Europe and Japan, fuel economy has traditionally been achieved through high fuel excise duty, but in the United States, the government has legislated for fuel efficiency. Corporate Average Fuel Economy (CAFE) was introduced in 1978 in response to the 1973/4 energy crisis and it set a maximum average fuel economy figure for all cars and later also light trucks, sold by a manufacturer during a given year. If the manufacturer did not reach this figure, a fine had to be paid, although it could buy credits from a manufacturer that performed better than the standard. CAFE limits were tightened up regularly until 1990 with little change thereafter. From 2011, a new system was introduced based on a vehicle's 'footprint', which is calculated by multiplying its wheelbase with its average track width. Larger cars are allowed to consume more than smaller cars, but it adds an element of sophistication to the system, albeit that manufacturers may be encouraged to enlarge their vehicles in order to optimize them for a given class or indeed move them into a higher class. The target is to achieve 250 g/km of CO_2 equivalent by 2016 under the US CAFE test cycle, which equates to 172 g/km under the European NEDC cycle (ICCT, 2011).

14.5 Climate Change

Regulation of the car was prompted by air pollution resulting from its emissions of toxins, as these were found to be injurious to human health. With climate change, however, the situation is different. This is far more controversial in that the scientific consensus is less obvious both in terms of the mechanisms by which CO_2 emissions are linked to climate change and the relative importance of the automotive contribution to that process. Also, there is no clear and imminent danger to human life or health; at least not one that is easy to demonstrate, as we are dealing with processes that are well beyond 'human measure', often playing out over long timescales (Kroonenberg, 2006; Stager, 2011; Nieuwenhuis, 2014). This has enabled the automotive industry to adopt a much stronger lobbying position. However, in reality, it is the same as legislating for fuel economy, so need not be so controversial.

After experimenting with self-regulation, the EU took the lead in reducing CO_2 emissions from cars. In April 2009, the EU adopted Regulation 443/2009, which established a CO_2 emissions target of 130 g/km for the average of new cars sold by 2015, with a limit of 95 g/km for 2020, although this is meeting strong opposition. In the West, the effects of the post-2008 recession indicated a consumer shift towards lower carbon vehicles, either as a primary choice or as an indirect result of the various scrappage schemes, which enticed consumers who would normally favour used cars into buying lower priced new cars, which tend to be smaller and more fuel efficient (Wells et al., 2010). This trend helped to accelerate the carbon reduction process in the EU. Other jurisdictions have also implemented CO_2 reduction policies as well as fuel efficiency policies, notably Japan, South Korea, China and the United States (see earlier text) as well as California.

China has less of an established parc (which means vehicles in use) and car buying habits can therefore still be changed through incentive schemes without too much of a legacy problem, although this window is closing. Decision making processes can be much quicker in China than in the EU with its 27 member states and complex multiple interest and lobby groups. Chinese authorities are better able to act unilaterally on the basis of environmental criteria alone (Nieuwenhuis and Zapata, 2005; Gallagher, 2006). The option of promoting smaller, more fuel efficient cars was among the first to be implemented (Zhengzheng, 2006). The central government issued a notice to encourage the use of environment-friendly, low-emission cars. This includes small cars, diesel powered vehicles and alternative fuel vehicles, known as 'new fuel' vehicles. To back up this policy, a number of incentives were introduced (Nieuwenhuis, 2012: see also chapter 10). The government also suggested reduced parking charges for small cars, as in Japan (see next paragraph) and announced that all restrictions on the use of small cars be lifted – Guangzhou, for example, banned cars of less than 1000 cm^3 capacity in 2001. Since then more fuel economy targets have been introduced, as well as targets for electric vehicles. The fuel economy targets issued in December 2009 aim to reduce fuel consumption of passenger vehicles to 7l/100 km, or 167 g/km of CO_2 by 2015 (ICCT, 2011).

In 1976, in the wake of the energy crisis of 1973–1974, Japan introduced the Energy Conservation Law under which various fuel economy standards have since been introduced. The Japanese market has long featured a midget car or 'kei' car segment. Vehicles in this segment have a maximum length of 3.4 m, a maximum width of 1.48 m and a maximum height of 2 m. Maximum engine capacity is 660 cm^3. In return, they enjoy a favourable regulatory regime. Kei cars tend to be exempt from the expensive urban parking permit requirement, for example. Since 1998, Japanese fuel economy regulation has been using the 'top runner' principle, whereby current best practice informs future targets. The targets introduced in 2007 were designed to lead to an average new car fuel economy of 16.8 km/l by 2015, equivalent to a CO_2 figure of 125 g/km, giving Japan the lowest CO_2 emitting new car fleet by 2015 (ICCT, 2011).

During the 1990s, the South Korean government also adopted a minicar policy. Here the principal driver was social equity and access to cars. The introduction of minicars was resisted by Korean manufacturers fearing small profits, but several of them have been successful with these products worldwide, for example Daewoo's Matiz. Korean regulation differentiated their minicars from the Japanese Kei model to avoid imports from Japan and promote indigenous designs; this has also helped make them more appealing in export markets. Korean minicars have a maximum engine capacity of 800 cm^3 and are slightly larger. However, South Koreans have increasingly favoured larger engined cars and SUVs, so in 2010 the government proposed a combined fuel economy and CO_2 emissions target of 17 km/l and 140 g/km of CO_2 on the Korean drive cycle for the 2015 model year. This equates to 150 g/km on the European NEDC drive cycle (ICCT, 2011).

As outlined earlier, California was the first to legislate to control vehicle emissions, predating federal initiatives, and for this reason retains the right to set its own standards separate from Washington. However, on climate change it was long hampered by federal policy with some innovative initiatives being challenged. In 2002, it enacted the first restrictions on GHG emissions from vehicles, which were made more specific in 2004 creating two vehicle classes and setting targets for a year on year reduction in emissions from 2009 to 2016. From 2009, the California Air Resources Board was able to align its measures more closely with federal standards. California therefore aims to achieve 172 g/km by 2016 but is aiming for tighter targets for model years 2017–2025. A target of 99 g/km is under study for 2025 (ICCT, 2011). Australia, which sets its own standards, has been lagging somewhat behind in this area due to various governments' attitudes to climate change. Under Prime Minister Kevin Rudd's administration, and to a lesser extent under Julia Gillard, the country was trying to catch up without unduly harming its indigenous car makers. In 2005, a voluntary target was set by the Federal Chamber for the Automotive Industries in Canberra for 222 g/km by 2010. Some real efforts were made under this initiative such as promotion of LPG and the tentative downsizing of engines for the traditional Australian cars. In addition, a greater diesel penetration was promoted (Nieuwenhuis, 2012). However, decisions by Ford and GM in 2013, followed by Toyota in 2014, to stop production of these indigenous models from 2016 and 2017 somewhat ease this problem.

14.6 Future Developments

Emissions reduction is all about energy efficiency – that is using less energy to do the same thing and yet is still often considered as a cost to business, a view challenged by, for example, Porter and van der Linde (1995). Clearly, technologies and methods have to be developed in order to save this money, but once in place, they will be paid for. Once in place, such technologies provide an advantage to those who developed them and can implement them first. They will become valuable in their own right and can be sold to those lagging behind. Perhaps a greater constraint on such technological developments is the expected imminent peaking of various commodities and specialist materials we use for such technologies (Heinberg, 2007).

The notion of oil peaking is increasingly well known (Roberts, 2004). The other commodities we need to cause concern include platinum, which is used for catalytic converters, as well as fuel cells, and lithium, which is needed for most of our electric vehicle batteries, as well as those used in laptops, mobile phones and so on. Half the world's lithium is in Bolivia, for example, while the so-called rare earths – such as the materials used for touch-screens – are controlled by China. At least the processing facilities are and not surprisingly, the Chinese have decided to keep most of it for their own industry, prompting the West to rediscover rare earth supplies in places like California and Finland and establish processing plants near those locations.

Yet it is clear that for the next 10 years, conventional petrol and diesel powertrain will continue to dominate the new car market. This is due to current product cycles, model replacements planned for the next few years and the fact that the most popular alterative powertrain – the hybrid – still uses an internal combustion (IC) engine. At the same time, oil-derived fuels are likely to increase in cost, albeit as part of a trajectory of considerable price volatility, which will see downs as well as ups. Supply of 'cheap' oil is estimated by some observers to peak around 2010–2015, while demand – from newly motorizing nations such as China, India, Indonesia and Russia – will continue to increase (Roberts, 2004; Heinberg, 2007; Hirsch et al., 2007; Kendall, 2008). However, the diesel/petrol mix can be as important to oil prices as the supply of crude oil itself. The move towards cleaner fuel by the shipping sector, for example, could potentially increase worldwide demand for diesel, competing with road vehicles.

This could boost demand for alternative powertrain technologies. Alternatively, the car industry could – in a bid to preserve the tried and trusted IC engine – go instead for enhanced conventional IC powertrain technologies such as petrol- or diesel-hybrid solutions to meet tightening emissions standards. Current hybrids still use petrol or diesel fuel, after all, unlike full electric powertrain, such as battery-electrics or fuel cells, which have the potential to make IC obsolete and as a result render worthless much of the car makers' investments in conventional IC technology. IC fuels can also be derived from natural gas (GTL), coal (CTL) or biomass even when oil itself becomes too costly (Nieuwenhuis, 1999; Kendall, 2008; Sperling and Gordon, 2009). This perpetuation of internal combustion could well be used to postpone the inevitable moment when internal combustion will itself no longer be viable.

References

Gallagher, K.S. (2006), 'Limits to Leapfrogging in energy technologies? Evidence from the Chinese automobile industry', *Energy Policy*, **34**: 383–394.

Heinberg, R. (2007), *Peak Everything; Waking up to the Century of Declines*, Gabriola Island: New Society.

Hirsch, R., Bexdek, R. and R. Wendling (2007), 'Peaking of world oil production and its mitigation', in Sperling, D. and J. Cannon (eds) *Driving Climate Change; Cutting Carbon from Transportation*, pp. 9–27, Burlington: Academic Press/Elsevier.

ICCT (2011), Global Light-duty Vehicles: Fuel Economy and Greenhouse Gas Emissions Standards, April 2011 Update, Washington, DC: International Council on Clean Transportation, www.theicct.org, (accessed 11 June 2011).

Kendall, G. (2008), *Plugged In; The End of the Oil Age*, Brussels: WWF.

Kroonenberg, S. (2006), *De Menselijke Maat; De aarde over tienduizend jaar [The Human Measure; The Earth in Ten Thousand Years Time]*, Amsterdam/Antwerp: Atlas.

Nieuwenhuis, P. (1999), 'The gaseous alternative', *Diesel Car*, **133**, October: 62–63.

Nieuwenhuis, P. (2012), 'The challenge of decarbonising the car', in M. Nilsson, K. Hillman, Rickne, A. and T. Magnusson (eds), *Paving the Road to Sustainable Transport: Governance and Innovation in Low-Carbon Vehicles*, pp. 17–39, London: Routledge.

Nieuwenhuis, P. (2014), *Sustainable Automobility; Understanding the Car as a Natural System*, Cheltenham: Edward Elgar.

Nieuwenhuis, P. and C. Zapata (2005), 'Can China reduce CO_2 emissions from cars?', *Greener Management International*, **50**(2005), Summer: 65–74.

Porter, M. and C. van der Linde (1995), 'Toward a new conception of the environment-competitiveness relationship', *Journal of Economic Perspectives*, **9**(4), Fall: 97–118.

Roberts, P. (2004), *The End of Oil; The Decline of the Petroleum Economy and the Rise of a New Energy Order*, London: Bloomsbury.

Sperling, D. and D. Gordon (2009), *Two Billion Cars; Driving Towards Sustainability*, Oxford: University Press.

Stager, C. (2011), *Deep Future; The Next 100,000 Years of Life on Earth*, Toronto: HarperCollins.

Wells, P. (2007), 'Deaths and injuries from car accidents: An intractable problem?', *Journal of Cleaner Production*, **15**(11/12), 116–121.

Wells, P. (2010), *The Automotive Industry in an Era of Eco-Austerity; Creating an Industry as if the Planet Mattered*, Cheltenham/Northampton: Edward Elgar.

Wells, P., P. Nieuwenhuis, Nash, H. and L. Frater (2010), *Lowering the Bar: Options for the Automotive Industry to Achieve 80g/km CO2 by 2010 in Europe*, Cardiff: CAIR/BRASS for Greenpeace International.

Zhengzheng, G. (2006), 'Green light given to eco-friendly vehicles', *China Daily*, 5 January, www.chinadaily.com.cn/english/doc/2006-01/05/content_509279.htm, (accessed 16 August 2013).

15

Global versus Local: Regionalism in a Global Industry

Paul Nieuwenhuis[1] and Peter Wells[2]
[1] *Centre for Automotive Industry Research and Electric Vehicle Centre of Excellence, Cardiff Business School, Cardiff University, Cardiff, Wales, UK*
[2] *Centre for Automotive Industry Research, Cardiff Business School, Cardiff University, Cardiff, Wales, UK*

15.1 The Old World

As pointed out in Chapters 3 and 6, despite globalization, localism persists in car markets and industries and is unlikely to go away. Regulatory regimes, taxation systems, and cultural and historical factors play a role. Such differences continue to shape the industry and it is therefore worth exploring some examples in a bit more depth. In the established car-making and car-using countries of Europe and North America there remain distinct differences. Europe, the region that effectively invented the car, still tends to favour smaller, more agile and more fuel-efficient cars. North America, where mass car production was invented (see Chapter 5), still favours larger, less fuel-efficient vehicles, aimed more at comfort than agility.

In recent years, there has been a degree of convergence, with European cars becoming larger and heavier, while American cars staying roughly the same size and weight, but becoming more agile, more fuel efficient and more aimed at drivers than before. On the other hand, many of the traditional North American automotive values have moved into vehicles that are categorized locally as 'light trucks', which comprise SUVs, minivans and pick-up trucks. In terms of product, then, there is still a difference, yet Americans also buy Asian and European products, creating a degree of diversity in the market, while Europeans also buy Asian cars and buy few American cars, yet also increasingly adopting SUVs, and minivans, albeit that these are smaller and lighter on average than their equivalents in North America. Other differences involve regulation on emissions and safety, which have to some extent been covered in Chapter 14.

The Global Automotive Industry, First Edition. Edited by Paul Nieuwenhuis and Peter Wells.

So, why do the United States and Canada favour larger cars and light trucks? The answer has often been sought in the longer distances travelled, making comfort of prime importance, although the straightness of the roads has often been exaggerated; the many mountain ranges feature as many winding roads as can be found in Europe. More important has been the price of oil and the fact that the United States in particular was self-sufficient in oil during the formative phase of automobility development (Nieuwenhuis, 2014). By contrast, Europe's culture of light, agile and fuel efficient cars is due to the fact that few countries in Europe were self sufficient in oil while they adopted motoring, and fuel tax has long been used by governments to control demand and protect the balance of payments with oil rich countries. As a result, fuel efficiency has been central. In addition, motor sport developed differently in Europe leading to cars that, while far less powerful than those across the Atlantic, were nevertheless often designed for a particular set of performance parameters, which optimized the performance of smaller engines and capitalized on the greater agility possible with lighter cars. This also made European cars more fragile, as they were developed to much tighter 'performance envelopes', making them less forgiving and generally less robust. Also, whilst in much of Europe a small garage or independent mechanic was never too far away when one did break down, a driver would not want to be stranded in the middle of Death Valley, or a Michigan winter and American cars were therefore designed with much greater reserves, through great simplicity and large, unstressed engines. At the same time, the Detroit automakers were virtually alone in achieving genuine economies of scale and therefore their products were genuinely mass-produced, often down to a price. Few cars in Europe managed this feat, with the exception of cars like VWs Beetle. Even today, few European models achieve genuine economies of scale (Nieuwenhuis and Wells, 2009). This combination of characteristics also led to a situation whereby luxury cars, largely phased out in the United States in the wake of mass production, continued to be viable, with higher price levels compensating for higher engineering content and greater luxury through manual input. Today, this makes Europe's industry unique in being able to offer prestige products with global appeal, such as those made by BMW, Mercedes-Benz, Jaguar, Bentley, Maserati and others.

American luxury cars, as typified by General Motors' Cadillac brand, became mass produced, although using higher quality materials and made in smaller numbers than more mainstream brands in the same group, while also deriving some of their value from additional equipment, often highlighting technologies such as cruise control that were adopted in Europe decades later. Yet these products were far from being drivers' cars and they increasingly lost out to better handling European and Asian imports. More recently, US carmakers, including Cadillac with its CTS and ATS, have started to emulate some of these imports and are catching up in terms of drivability, yet retaining their mass production cost advantage. In the case of Chrysler, its 300 and Crossfire, as well as their related Dodge models, benefited directly from using European platforms while under Daimler ownership. Chrysler products subsequently started to adopt FIAT technology for some products. However, the size of the US market and the resulting achievement of these economies of scale make exports of North American products largely unnecessary. Despite limited sales of Chryslers in Europe and even smaller numbers of Cadillacs and Corvettes, these all sell well in North America. On the other hand, China has for some time been a good market for a range of Buick products developed in the United States and Australia, prompting GM to at least retain the Buick brand worldwide, despite its eroded brand value in North America.

15.2 Asia

Within Asia, there are also considerable national differences in automotive cultures. The rise of Japan's car industry also followed a fuel efficiency path, as in Europe, favouring even smaller cars, such as the unique 'kei' or midget car segment, which are on the whole confined to the home market, where they are not always regarded as proper cars. The other cars made here do combine European style fuel efficiency with high-build quality and great reliability, which have made them popular around the world for day-to-day motoring. A similar model has been followed by the Korean automotive industry which, partly perhaps through key global events such as the foreign takeover of Daewoo, has come to recognize that European vehicle attributes in terms of performance, packaging, handling and styling are applicable to and successful in many markets. Korean car companies therefore hired many Europeans to help them improve in these key areas and it, combined with the inherent quality and efficiency of their products, has allowed them to make major inroads in all key markets, at the expense of Japanese, European and American manufacturers. Within Korea, however, a unique motoring culture also exists, with few imports and with local manufacturers offering a full range of models, covering most segments with products often not available in export markets; Koreans mostly still favour local products.

In the 1980s, other Asian countries also embarked on car making ventures, but why are those leading this thrust, notably Taiwan and Malaysia, marginal players today? Taiwan soon realized that it would be unable to compete with Japan and Korea in car making, although some local assembly continues. Instead, with rising wages and rising technological expertise, it ventured into higher technology areas in ICT and aerospace. Malaysia's car industry was set up with intensive state support and still enjoys a key position in the home market. However, it still relies heavily on outside input and has not managed to build the momentum and scale to challenge its rivals on world markets. Neighbouring Thailand on the other hand has become a significant car assembler, albeit of foreign designs with facilities largely foreign owned or run by local licensees. However, this large local assembly sector does provide many firms with important access to ASEAN markets in the region, while certain products are also exported to world markets. Other countries that may become players in future include Iran and Indonesia, both already significant markets with some local production and assembly.

15.2.1 The Creation of Two Motoring Cultures: India v China

India and China are covered in detail in Chapters 11 and 10 respectively, but some additional points can be made in the present context. After independence, India adopted a very protectionist model for its car industry. Car imports were effectively banned, or subject to punitive import duties, designed to allow a native industry to develop. For many years, this industry relied on a few dated, foreign designs, notably the legendary Hindustan Ambassador (Morris Oxford), Premier Padmini (Fiat 1100) and Standard Herald (Triumph Herald). The large Contessa (Vauxhall Victor) was added to the Hindustan range in 1983. Heavy truck makers relied on Mercedes (TATA) or Leyland (Ashok) designs, while light commercial vehicle designs were adopted from Germany (Tempo). This mix served the country's basic motoring needs for many years; however, in the 1980s, the government added to this mix through its Maruti joint venture with Suzuki. India now had a small, cheap car and Maruti began the motorization of the Indian middle classes who began to benefit from a growing economy. India's economy was gradually opened up from the late 1980s onwards and with this came new opportunities for vehicle producers such as Ford, Daewoo, GM, Fiat, Skoda, Honda and Hyundai. India has also begun to develop a niche in small electric cars, notably Reva of Bangalore, which enjoys significant exports, prompting a sale to Mahindra (Nieuwenhuis, 2014). Importantly, the Maruti largely defined customer expectations of what constitutes a car and as a result, India has essentially developed as a small-car market.

By contrast, as outlined in Chapter 10, China's car industry started in 1952 and was very limited for many years with two large cars aimed at senior party officials: the Mercedes-inspired Shanghai and the FAW Hong Qi (Red Flag). Trucks played a more important role, with Dong Feng's medium-duty trucks particularly important. Cars were restricted to various forms of official use. However, some early joint ventures began to break the monopoly of dated-local designs. Chrysler-Jeep (Beijing) and Volkswagen (SAIC and FAW) were early entrants, followed later by other foreign firms, including General Motors, Iveco, PSA and Toyota-Daihatsu with its popular Xiali. Until recently, Chinese production was primarily of cars designed abroad. In recent years, this has begun to change, although many designs are still derivative of Western or Japanese designs, while others actively engage foreign design and engineering expertise to develop unique cars to international standards. China is now the world's largest car producer, primarily for a rapidly growing domestic market. What is without doubt, however, is the ability of China's car industry to build cars in high volumes and to increasingly higher standards.

Nevertheless, exports are limited at present, which with the still rapidly growing home market is not yet a problem. As the expectations of the home market develop, quality standards will improve, while increasingly, local solutions and designs will emerge. One significant step in this direction is BYD's lead in electric cars, stemming from the fact that its main business is making batteries. Moves such as Geely's takeover of Volvo also forge stronger links with foreign automotive technology and design leaders. However, government at both national and regional level continues to play a significant role in the industry. Unlike India, China has developed as a larger car market. The expectations of what a car was were based on the cars built for official use in the early days of motorization and these were largely the VW Passat, Audi 100 and Buick. This format has become embedded in customer expectations in this market, leading even to some local authorities banning small cars as being unsafe (see Chapters 10 and 14).

15.3 Latin America

As South America's largest country, Brazil was keen to develop an automotive industry. However, there was an awareness that there was limited local expertise. For this reason, from the 1950s, the government embarked on a policy of attracting inward investment by foreign car companies, but insisted on high levels of local content which encouraged a home-grown automotive supply sector to develop (Nieuwenhuis, 1988). It was hoped that in time this would allow the development of local car manufacturers. This dream did not quite come true, although a number of small local producers did emerge, such as Gurgel, Puma and Agrale. By the 1960s, the local car industry was dominated by Volkswagen, DKW-VEMAG, a German-Brazilian combine boasting some local designs, Simca (France/US), Willys-Overland (American Motors) and Alfa Romeo. Later, Toyota, GM and Chrysler played a role, Ford made a Renault-derived car, while Fiat, a late entrant in 1988, rapidly became a dominant player in the 1990s. Brazil developed some local designs, some of which (Puma, Volkswagen SP2 and VW Brasilia) enjoyed some exports (Nieuwenhuis, 1988). However, as globalization progressed, Brazil became part of the global networks of several manufacturers, notably Fiat and Volkswagen. Fiat's global network of production for newly motorizing markets – spearheaded by the Palio model – was centred on Brazil, while China enjoyed a Brazilian version of the VW Passat as one of the key models during its fastest growth phase. The truck industry is dominated by Swedish firms Volvo and Scania, with VW a secondary player.

One special contribution of Brazil to automotive technology is its expertise in bioethanol. Under a programme started in the 1970s, the government promoted the use of bioethanol as a substitute for expensive imported oil. It made the sugar, oil and car industries cooperate and by the mid 1980s the programme could be considered a success. However, falling oil prices soon made ethanol uncompetitive and the Proalcool project declined. At this stage, cars had to be built to use either petrol or ethanol. However, a breakthrough came when local branches of suppliers Bosch and Magneti Marelli developed 'flex-fuel' technology in the early years of this century. Cars could now sense which fuel was being used and adjust the engine accordingly. At the same time, with rising oil prices, ethanol enjoyed new appeal and today the vast majority of Brazilian cars are capable of running on ethanol, or any mixture of ethanol and petrol. For the consumer, the advantage is also that cars perform better on ethanol. The technology and the fuel are now both exported to Europe and North America (Zapata and Nieuwenhuis, 2009).

Elsewhere in Latin America, significant assembly facilities developed, notably in Argentina and Mexico. Especially the latter, with close links with the United States, and its incorporation into NAFTA, became a source of lower labour cost manufacturing for the more developed North American markets, as well as enabling exports of American and European products elsewhere. However, neither has yet managed to create a significant indigenous industry.

15.4 Case Study: On the Margins of Mass Production: Australia

The Australian car market is an interesting one. Though subject to many of the globalizing trends now shared by virtually all markets, it has managed to retain one unique local segment: full-size rear-wheel drive cars, estates and the famous 'utes' derived from these. Car making has come under increasing pressure in recent years, particularly from an appreciating Australian dollar, as governments prioritized short-term gains from mining to building a longer term resilience in manufacturing, as Norway has tried to do with its mineral wealth. This culminated in the decision by both Ford and GM in 2013 to cease manufacturing by 2017, and Toyota announced in 2014 that it was also to cease manufacturing (Nieuwenhuis, 2014).

This market was long served by local branches of Ford and General Motors (Holden), but Toyota and Mitsubishi who for many years stuck with global designs, also decided to build locally-designed large cars in Australia, albeit with FWD, thereby creating their own niche. Mitsubishi offered the 380 in this niche and Toyota the Aurion. In February 2008, Mitsubishi announced the closure of its Adelaide plant and the phasing out of the 380. The existence of the large RWD segment allowed local manufacturing to survive, but Ford and GM have had to be very clever about how to go about this. Success was achieved through a unique interplay between the manufacturing systems and the markets they serve.

The Australian full-size segment features a traditional rear-wheel drive and front-engine layout in a car that is large by European standards and midsize by North American standards, but with chassis characteristics and trim quality that are more European than American. As well as traditional estate derivatives, these platforms always feature a 'ute', or car-derived pick-up truck in their lineup, sometimes complemented with a chassis–cab variant, while Ford offered a SUV derivative in recent years, the Territory. The segment survived and developed in virtual isolation with some exports to Asian and Middle Eastern markets for Holden, and New Zealand for both Ford and Holden. Holden also sent some cars to the United Kingdom, while Ford exported cars to the United Kingdom for the funeral trade, marketed by Coleman Milne.

By the late 1990s, General Motors had shifted completely to front-wheel drive platforms in all major markets. Even the last European rear-wheel drive platform, Senator, had been phased out. However, the global environment was changing; not only had the German specialists and rival Ford's Jaguar stuck with rear-wheel drive for their prestige platforms, but an increasing number of direct competitors were shifting to rear-wheel drive: notably Lexus, Infinity and even Chrysler (using Mercedes platforms, see earlier text). GM was being left behind, except that Holden had stuck with the large rear-wheel drive platform. GM was able to leverage their expertise and commissioned the Australian division to develop a new world class rear-wheel drive platform – Holden launched its VE Commodore, developed at a cost of over a billion Australian dollars (US$ 780 million; €606 million), in 2006. The platform is used for cars in various divisions in the global GM empire, including Buick and Daewoo in Asia, Chevrolet in North America (including Camaro), and it is related to the RWD platform used by Cadillac, while some sales are achieved in the United Kingdom. At the 2007 Chicago Auto Show, Pontiac showed the G8, a modified VE Commodore, available in the United States from early 2008, while Vauxhall marketed a version of the car as the VXR8 in the United Kingdom. Earlier experiments to market the Monaro Coupe in the United Kingdom (Vauxhall Monaro) and United States (Pontiac GTO) led to limited sales but established the feasibility of the manufacturing model. The G8 project was scuppered by the closure of GM's Pontiac division as part of GM's post bail-out restructuring.

Australia's car industry is one that deserves particular attention for a number of reasons. It is driven by unique local cultural and market expectations, while in manufacturing terms, it uses conventional Ford-Budd mass production systems and technologies based on conventional all-steel body designs, yet operates at the margins of viable economies of scale for such technologies. In fact, Holden's first car, launched in 1948, was based on GM's first US monocoque design, initially intended for use in the United States where a downsizing trend was expected after the war, but when this failed to materialize, it was offered to Holden (Wright, 2008).

By 2014, three firms still operated mass production facilities in Australia – General Motors Holden, Ford Motor Company of Australia and Toyota Motor Australia. Mitsubishi Motors Australia Ltd. closed its factory in 2008. Production is centred in two locations: Melbourne (Ford, Toyota) and Adelaide (Holden, Mitsubishi – closed in 2008). Though headquartered at Fishermen's Bend in Melbourne, Victoria, Holden production takes place at the Elizabeth facility outside Adelaide in South Australia. In 2004–2005, the plant enjoyed a comprehensive refit in preparation for the new VE. This was achieved while the plant continued to operate and at record levels for some of the period. This process illustrates some of the lean practices that enabled the Australian facilities to compete for such a long time.

Another important factor is that these divisions are able to breathe in tune with the local market. Their marketing divisions are able to respond quickly to subtle changes in local requirements. The focus on one primary market makes this possible. Toyota and Mitsubishi have long been less adept at this game, producing much more globally-informed variants. Ford and Holden, however, offer uniquely Australian products, which is clear from the colour choices offered and in fact chosen by customers. Colours that appear unsuited to large cars in more conservative markets find a ready market in Australia and New Zealand, from bright orange to light metallic green. Interiors are similarly bold and Ford created a special in-house facility at its Broadmeadows plant to produce these striking interior colour combinations, enabling the firm to respond in a matter of weeks to subtle changes in colour preference from the market without relying on negotiations with external suppliers.

Economies of scale were also achieved by building a number of closely related derivatives of the main platform on the same production lines. While the saloon is the core product, there is always a predictable demand for estates and utes. Each of Ford and Holden's main platforms also has a longer wheelbase derivative for the luxury variants, such as the Holden Statesman, and its sportier Caprice version, and Ford's Fairlane and Ltd. In addition, Holden revived the Monaro Coupe in the early 2000s on the same platform. As mentioned earlier, it even exported this variant to the United Kingdom, as well as the United States, where it was badged Pontiac GTO, although sales at a peak of 13 589 in 2004 were disappointing – conservative styling was blamed. Similarly, Ford created the Territory SUV, which used a number of Falcon modules, whilst looking very different from the saloon. Engines included a locally assembled in-line 6 (Ford – later dropped in favour of a global V6), V6 (Holden – supplied to other GM plants) and US-sourced V8 units. More recently 4-cylinder units were also offered in some variants. Local manufacturers complete their model ranges with imported models, using their parents' global networks as a source. Such vehicles are adapted to local conditions and – in the case of GM – wear a Holden badge. Both firms also offer a range of high performance variants of these vehicles in all body styles, which are highly regarded in the local market. Holden created HSV (Holden Special Vehicles) in partnership with TWR of the United Kingdom, while Ford operated the FPV (Ford Performance Vehicles) division in collaboration with Tickford in the United Kingdom.

Of course the fact that all local players are part of global corporations is important and ensured their survival. Yet, at the same time, all had to earn their keep within those corporations. Holden has traditionally adopted one of GM's platforms, usually a European one, and then, through a series of facelifts and model changes adapted the platform to such an extent that little commonality remained with its original source. However, because of the proportion of adopted and carry-over components, costs have been kept down. For each succeeding model generation, a limited number of components and subassemblies are changed, thus keeping investment and manufacturing costs within manageable limits for the volumes produced. Yet, over time, considerable localization is achieved. Thus, while late 1990s Holden cars retained a hint of slightly oversized Opel models, they had in fact become very different, having been fine-tuned to Australian market expectations and conditions.

Antipodean drivers expect to be able to use any car from the cool and damp conditions in Tasmania or South Island New Zealand to the dry and dusty deserts of the Northern Territory, from the motorways of Sydney and Melbourne to the gravel roads and rutted tracks of the outback, while negotiating the occasional ford through a creek in the Flinders Ranges. Where the average European or even US driver would baulk at the conditions, or slow down to walking pace, Australians would expect to keep up a decent 100km/h on outback gravel roads (Nieuwenhuis, 2014). Cars designed and built for this range of conditions do have credibility in other markets, and the Middle East became a key export market for Holden, to the extent that a round of the uniquely Antipodean V8 Super Cars racing series came to be run in the Gulf. Motor racing has long been used as a marketing tool by the local manufacturers, with the annual Bathurst 1000km race the highlight of the season. The industry has managed to build fierce loyalties among fans with Ford and Holden fans bitterly divided. In Australia you are either a Holden supporter or a Ford fan, a division that may soon become global, if Australian exports continue their upward trend (Bedwell, 2009).

With the increasing pressure from diversifying markets on the ageing mass production system, the world's mass car producers may well be forced to adopt the flexibility of the Australian producers in order to survive. In the past, Australia consistently implemented policies to protect its indigenous car sector, for example, the then Prime Minister Kevin Rudd announced in 2008 an AUS$6.2 billion (US$4.3 billion) plan to make the country's automotive industry more economically and environmentally sustainable by 2020. The plan featured an expanded AUS$1.3 billion Green Car Innovation Fund, which provided government funding for companies to design and sell environment-friendly cars. The Australian government would match industry investment in green cars on a AUS$1-to-AUS$3 basis over a 10-year period from 2009. The 'New Car Plan for a Greener Future' was expected to generate AUS$16 billion in investment in the Australian automotive industry over the 13-year life of the project.

The government also announced that automotive tariffs would be cut to 5%, giving Australia the third-lowest tariff regime among economies with a well-developed automotive industry. Unfortunately, events overtook this initiative, with the Gillard and subsequent right-of-centre Abbott administration more inclined to support the extractive sector, which led to an increase in commodities dependence and – with rapidly increasing demand for these resources, especially from China, producing for world markets – it prompted a steady increase in value of the Australian dollar, rendering car making there effectively not economically viable, thereby prompting the decisions by Ford, GM and Toyota to pull out.

References

Bedwell, S. (2009), *Holden vs Ford; The Cars, The Culture, The Competition*, Dulwich Hill, NSW: Rockpool.

Nieuwenhuis, P. (1988), 'Brazil, the awakening giant?', *Motor Industry Review*, **3**, (88), 161–180.

Nieuwenhuis, P. (2014), *Australian Cars and the Industry that Made Them*, Leeds: Zeteo.

Nieuwenhuis, P. and P. Wells (2009), *Car Futures: Rethinking the Automotive Industry Beyond the American Model*, Cardiff, Wales: Trend Tracker, Ltd.

Wright, J. (2008), *Special; The Untold Story of Australia's Holden*, Crows Nest, NSW: Allen & Unwin.

Zapata, C. and P. Nieuwenhuis (2009), 'Driving on liquid sunshine – the Brazilian biofuel experience: a policy driven analysis', *Business Strategy and the Environment*, **18**, (8), 528–541.

16

The Impact of Electric Automobility

Huw Davies[1], Liana M. Cipcigan[1], Ceri Donovan[1], Daniel Newman[2] and Paul Nieuwenhuis[3]
[1]*Electric Vehicle Centre of Excellence, School of Engineering, Cardiff University, Cardiff, Wales, UK*
[2]*Electric Vehicle Centre of Excellence and Sustainable Places Research Institute, Cardiff University, Cardiff, Wales, UK*
[3]*Centre for Automotive Industry Research and Electric Vehicle Centre of Excellence, Cardiff Business School, Cardiff University, Cardiff, Wales, UK*

16.1 Electric Vehicle Design

The electric vehicle (EV) is by no means a 'novel' technology. The first electric car was introduced in 1834 by Thomas Davenport, thereby preceding the introduction of the first internal combustion engine (ICE) cars by Benz and Daimler by more than 50 years. With the addition of range-extended EVs, pioneered by, among others, Kriéger in France, and even a young Ferdinand Porsche whilst working as chief engineer for Lohner in Austria (Helmers and Marx, 2012), the electric car held the majority share of powered passenger vehicles by the start of the twentieth century (Høyer, 2008). However, the move to mass production of the ICE car pioneered by Henry Ford, coupled with the introduction in 1912 of the self-starter by Cadillac, led to the electric car being marginalized. Yet, interest in the electric car has grown in recent times. Expectations are that the development of different drivetrain technologies and charging possibilities will help accelerate the uptake and development of the electric car.

The Global Automotive Industry, First Edition. Edited by Paul Nieuwenhuis and Peter Wells.

16.1.1 Battery Electric Vehicles

The simplest solution to electrification of the drivetrain is to remove the ICE. Battery electric vehicles (BEVs) operate on electricity only and the main components are primarily a high voltage (HV) battery, an electric motor (either alternating current (AC) or direct current (DC)) and a motor controller. The electric motor is able to deliver a constant and high torque over a broad range of speeds and thus most BEVs do not need a reduction gearbox; therefore, the complete drivetrain is less complex compared to conventional vehicles with ICE.

The BEV can also be more cost efficient than comparable ICEs. The BEV differs from the ICE car as it does not require additional systems such as a starter motor, exhaust or gearbox. A study by the *Institut für Automobilwirtschaft* (IFA) found that electric cars could cost up to 35% less to maintain than an average ICE (ENEVATE, 2013). Whilst for energy consumption, the BEV is charged directly from the electricity grid, where charger efficiency can range between 60 and 90%. The motor itself is highly efficient, the efficiency varying between 85 and 95% across the entire range of speeds. The conversion from battery DC power to AC for the motor by a modern inverter is possible with around 95% efficiency. This gives an overall efficiency of greater than 70%. In comparison, the efficiency of an ICE vehicle is below 20% (Morrill, 2008), although some diesel engines are now approaching 40%.

The BEVs available today are also very similar to conventional cars in terms of design, comfort and safety. Most OEMs use existing platforms or derive the design of a BEV from existing models or technology. The limitation is the battery. In the interests of longevity, EV drivetrains will only use a portion of the full energy capacity of the battery, as determined by the maximum and minimum battery state of charge (SOC). For a BEV, this drivetrain will only operate in charge depleting (CD) mode. The CD mode is where the SOC is above the target SOC and the vehicle receives its power from the battery. The distance driven in CD mode is classified as the all electric range (AER). The AER for current generation BEVs is typically less than 200 km with a battery capacity of 15–25 kWh. An example of a current generation BEV is the Nissan Leaf. The Leaf uses a 24 kWh Li-ion battery to provide a range of 175 km (based on the New European Drive Cycle, NEDC) (Fabio et al., 2013).

16.1.2 Hybrid Electric Vehicles

An intermediate step in electrification of the vehicle drivetrain is the hybrid electric vehicle (HEV). In its simplest configuration, the HEV does not plug into the electricity grid, rather it simply uses a small battery to improve the fuel consumption of a traditional ICE through the capture and reuse of energy from a process called regenerative braking. The reuse of electric energy is through an electric motor that is used to support the ICE during acceleration or in stop-and-go traffic. In order to drive serious distances in purely electric mode, vehicles need a significantly larger battery, so an extension of the HEV is the plug-in HEV (PHEV). The PHEV utilizes grid electricity and a battery with higher energy capacity enabling further distance to be travelled on all-electric power. The PHEV has the benefits of a BEV, such as higher efficiency and regenerative braking, but also inherits the merits from conventional ICE vehicles, such as a large driving range and fast refuelling.

The HEV can be further classified according to the manner in which power is delivered to the wheels, namely, a series, parallel or power split configuration. The range extended electric vehicle (REEV) is an example of a series configuration. The range extender consists of a small

ICE that is decoupled from the drivetrain, a fuel tank and a generator to produce enough electricity for the EV when the battery is depleted. The REEV has the benefits of a BEV, such as high efficiency and regenerative braking, but also inherits the merits from conventional ICE vehicles, such as a large driving range and fast refuelling. The use of this technology is beneficial over the conventional ICE vehicles, as this mechanical decoupling of engine and wheels means that the engine can be operated in an optimal region for maximum efficiency and achieve greater fuel economy. An example of a REEV is the GM vehicle, marketed as the Vauxhall/Opel Ampera in Europe, Holden Volt in Australia and Chevrolet Volt/Cadillac ELR in the United States. Like the Leaf, the Ampera also uses a Li-ion battery, but its capacity of 16 kWh is less than that of the Leaf due to the addition of an on-board ICE. A PHEV or REEV can operate in both the CD mode and charge sustaining (CS) mode. The PHEV or REEV will operate in CD mode using batteries only, allowing short trips to be run solely off electricity. Once the SOC reaches the minimum threshold or target SOC, the REEV will then operate in CS mode and the ICE will provide the propulsion but still in the form of electricity for the REEV. The Ampera/Volt REEV will first operate using battery only (range of 50–80 km) and then switch to the extended range mode using electricity generated by the engine (up to 500 km).

16.2 Charging Infrastructure – UK Case Study

Electric mobility offers an opportunity for diversification of the primary energy sources used in transport but also brings new risks, technological challenges and commercial imperatives (Cipcigan et al., 2012). The innovation captured by EV is the reduction of CO_2 emissions and the resulting revolution in road transport and also the total absence of toxic emissions, which is of particular benefit in urban areas.

Future EV demand forecast is difficult to predict and there is no consensus on the number of EVs and PHEVs in use by 2030. According to the UK Committee on Climate Change, in the United Kingdom, the high EV uptake pathway targets are that EVs, including BEVs and PHEVS, cars and vans, capture a 16% share of new vehicle sales in 2020 and 60% share by 2030 (Element Energy, 2013). These targets can be achieved via the support of the plug-in car grant (£300M), charging station places programme (£30M), low carbon network fund (annual allocation of up to £64M) as well as through favourable tax regimes, including plug-in vehicles receiving vehicle excise duty and company car tax exemptions, as well as enhanced capital allowances (Office for Low Emission Vehicles, 2011; Tsang et al., 2012).

The development of an EV market is dependent on the parallel development of the recharging infrastructure, which will result in a spatially uneven increase in electricity demand. Based on the location, the charging infrastructure is classified as residential (private area with private access) or commercial (public area with public access and private area with public access) (Ferdowsi et al., 2010; Raab et al., 2011). Public charging infrastructures, including central charging stations for large car parks and street charging stations, are starting to be installed in large numbers to help the deployment of EVs and address the customer's 'range anxiety'. It is anticipated that the first wave of EV owners will be concentrated in major cities.

At the time of writing, the market for EV charging infrastructure is still in the early stages of development in the United Kingdom and is dependent on government funding via schemes such as Plugged in Places and is centred on lead regions (OLEV Plugged in Places projects

(Office for Low Emission Vehicles, 2011)). A number of EV charger providers are promoting new financial models for charging infrastructure development and are contributing at all stages of the supply chain. Some examples are the Chargemaster's POLAR network, British Gas for home-charging points and Ecotricity 'Electric Highway' motorway charging network.

However, is charging an EV really as simple as plugging it in? Electricity distribution networks are typically designed for specific load carrying capability based on typical load consumption patterns. Battery charging of EVs will increase the power demand in distribution networks. EVs, seen as mobile loads, will have to plug into the electricity grid, but not necessarily at the same location and time on a regular, predictable basis. Depending on the location and the time the EVs are plugged in, they could cause local or regional constraints on the grid as distribution networks are currently operated close to their capacity. The operating capacity of a distribution network is limited by the transformer loading capacity, cable current carrying capacity and voltage boundaries (Papadopoulos et al., 2012). An infrastructure of intelligent, networked charging stations that are conveniently located, safe to operate and simple to manage will avoid network reinforcement. However, even small clusters of uncontrolled vehicles charging at peak periods could significantly stress the grid distribution system, slowing EV adoption and requiring major infrastructure investments (Cipcigan et al., 2012).

EV charging regimes depend on the control strategies that are implemented in the system (Ferdowsi et al., 2010) such as:

- An uncontrolled charging regime, which considers that the charging would be completely uncontrolled and commuters would start the EV's charging procedure as soon as they return home. This charging mode is considered as the worst-case scenario for the power system as it is likely that the charging coincides with existing peaks in demand.
- A dual tariff regime involves an incentive for users to charge the EV overnight, taking advantage of night tariffs. Different electricity rates, price plans and hours of night charges exist, depending on the utility that operates in each area. This charging regime is sometimes called delayed charging, considering delaying charging until the demand curve starts decreasing after the daily peak.
- A smart charging regime uses demand management techniques. In many applications, an 'aggregator' entity is managing the existing infrastructure avoiding network reinforcement. This strategy is in line with 'Smart Grids' principles.

Research results show that an uncontrolled EV charging regime will increase the British winter day peak demand by 3.2 GW (3.1%) for a low EV uptake case (7%) and by 37 GW (59.6%) for a high EV uptake case (48.5%), in 2030. The British predicted energy demand for uncontrolled charging for winter and summer seasons is presented in Figure 16.1 (Papadopoulos et al., 2011).

This research concluded that a low uptake of EVs in the United Kingdom (7%) is not enough to make a real impact on the energy demand at a national level. So, additional capacity will not likely be required over the medium term, while the uptake of EVs remains low. Previous research indicates that if the charging load can be matched to periods of low demand, the existing grid will support a significant number of plug-in vehicles (Papadopoulos et al., 2013).

Generally, individual EVs have relatively small power capabilities of the order of tens of kW. Many studies agree that the EVs visibility to the system operation and participation in the

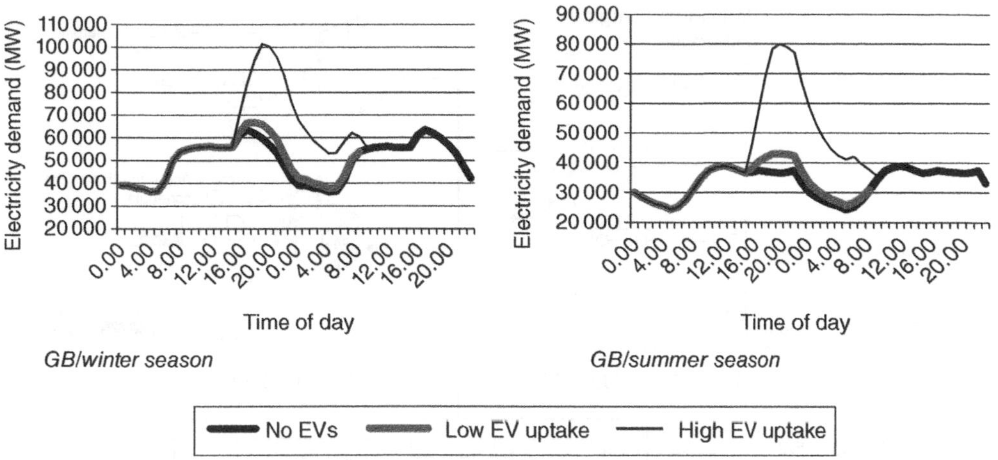

Figure 16.1 British predicted energy demand for uncontrolled charging in 2030. Source: EVCE, Cardiff University

electricity markets will require an 'aggregator', a new entity which enables their operation to be coordinated. The aggregator entity is a new player and will serve as an intermediary between large numbers of EVs, market players and/or system operators (Raab et al., 2011). The role of this entity is to cluster geographically dispersed resources and manage their generation and demand portfolios as a single entity. An EV aggregator is responsible for grouping the charging demand of a number of EVs in a residential area (residential cluster) and a number of EVs in a commercial area (commercial cluster). This new business entity is offering demand side management services to the market, managing both the aggregated groups of EVs (residential and commercial) by purchasing electricity from the power market (based on the load forecasting analysis) and selling it to the users. The aggregator is responsible for all necessary market activities for EVs: day-ahead demand forecast, contracts with the energy supplier, communication with the Distribution Network Operator (DNO)/ Distribution System Operator (DSO) and the Transmission System Operator (TSO). The aggregator is the entity responsible for both selling and contracting electricity to the users and aggregating their charging demand. To achieve these functionalities, the aggregator needs to exchange information with different entities through Information and Communication Technologies. The aggregator control centre makes decisions regarding any necessary adjustment in power exchange between its resources and the grid or to respond to some request signals of the system operator for system ancillary services (e.g. frequency control and voltage control). The EV aggregator's basic functions are presented schematically in Figure 16.2.

EVs represent a new type of non-linear load on the distribution system that could produce significant impacts on localized distribution systems. Distribution network infrastructure has different carrying capacities, and predicting the location and characteristics of the EV loads is important in the process of managing the load. Customer behaviour is not directly observable but may be understood based on past EV usage patterns. This type of load cannot be characterized by examining any previous comparable load on the system because of its spatial and temporal nature. In order to prepare the infrastructure to serve the anticipated electrical load, it is necessary to make predictions about where and when the load will be spatially located.

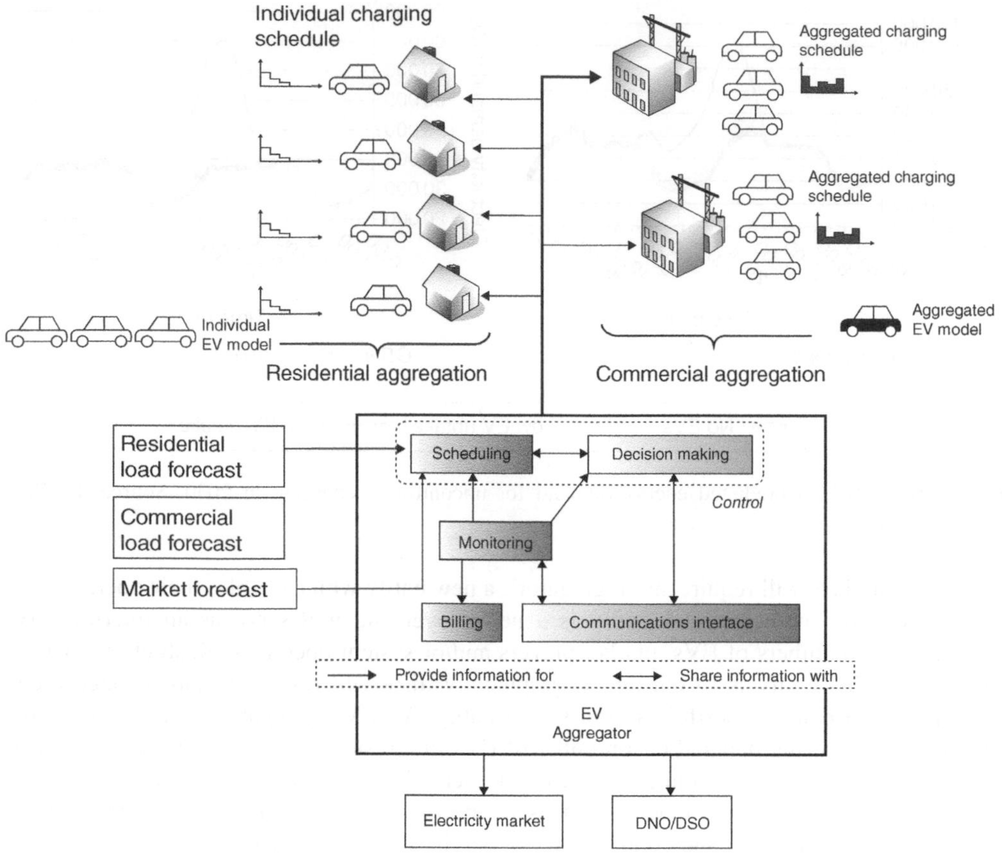

Figure 16.2 Electric vehicle aggregator basic functions

EVs are a mobile source of demand, charged for relatively long periods of time, and as a result, EVs could place significant coincident demand on the system. Forecasting EV penetration relies on predicting consumer behaviour in terms of purchasing and usage (Xydas et al., 2013a, b). At the same time, EV loads appear on the system without advance notice and in multiple locations and because of this random nature, load forecasting is an important issue.

The EV aggregator has to forecast the electricity consumption of its own customers (for forecasting the aggregator's power balance) or the consumption in the electricity system (for forecasting spot market electricity prices). Consumption can be forecast within various time horizons, and regarding the EV aggregator management, the most relevant is the short-term load forecasting, which considers time periods of up to one week. The main factors influencing the EV load are the time of the day, week or month, electricity price and customer behaviour (Xydas et al., 2013a). Customer behaviour is not directly observable but may be understood based on previous EV usage patterns. Such smart management of EV charging strategy based on aggregation could be considered a win–win strategy for both the DNO/DSO and the vehicle owner. It has been demonstrated that a control system using an EV supplier/aggregator is capable of managing EV battery charging within the technical constraints of the distribution network, whilst satisfying EV load demand during the cheapest hours of the day (Papadopoulos et al., 2013). But the future of EV aggregators will depend on the regulatory framework in each country.

16.3 Electric Vehicles in Europe

Although EVs are not a new concept, until recently, they have struggled to gain a significant market share. They have been successful in niche applications such as milk delivery or for use indoors or city centre delivery, where their characteristics such as silent use and zero tailpipe emissions have become an obvious advantage over combustion engine vehicles. However, growing concerns over oil availability and security, along with the need to reduce carbon emissions and improve air quality, particularly in major cities, have meant a growing interest in electric propulsion in recent years. Government incentives in many EU member states and a growing range of vehicles available, including models from major manufacturers, have driven a growth in EV sales in many EU countries, some of which are shown in Figure 16.3.

The growth in sales from 2010 onwards is encouraging. However, when placed in the context of overall vehicles sales, the percentage of EVs overall is small. Table 16.1 shows that even in Norway, which has the highest percentage market share of EVs, electric vehicles constitute only 1.2% of total new vehicle registrations.

Research carried out under the ENEVATE project (www.enevate.eu) asked potential consumers what incentives might encourage them to buy an EV. The results as shown in Figure 16.4 indicate that their prime concerns were cost and charging – compared to a conventionally powered car, they are too expensive (Newman et al., 2012).

Many European governments have been offering incentives to encourage the uptake of EVs. The UK government offers a Plug-in Car Grant of 25% (up to a maximum of £5000) against the purchase of new EVs. In the period January 2011–September 2013, 5702 grants had been awarded

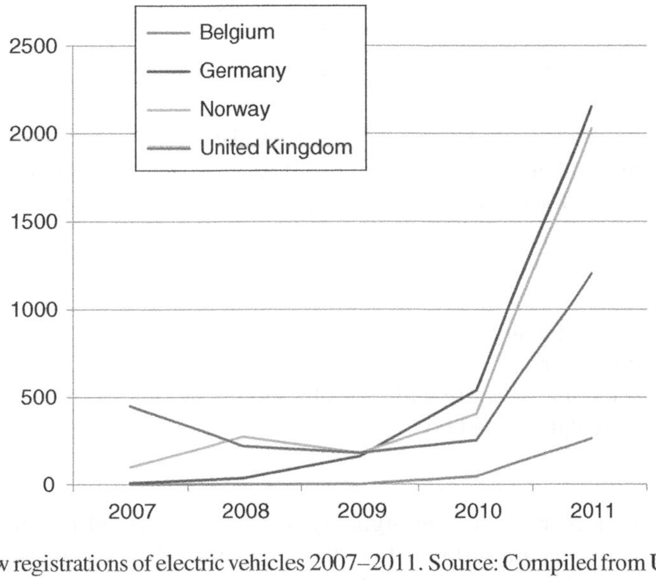

Figure 16.3 New registrations of electric vehicles 2007–2011. Source: Compiled from UNECE Statistical Database

Table 16.1 New vehicle registrations by propulsion type for 2011

	Petrol (units and % market share)	Diesel (units and % market share)	Electric (units and % market share)
Belgium	135 266 (23.69)	435 527 (76.27)	264 (0.05)
Germany	1 651 637 (52.44)	1 495 966 (47.49)	2154 (0.07)
Norway	37 924 (22.51)	128 559 (76.29)	2029 (1.20)
United Kingdom	924 509 (49.07)	958 536 (50.87)	1204 (0.06)

UNECE Statistical Database.

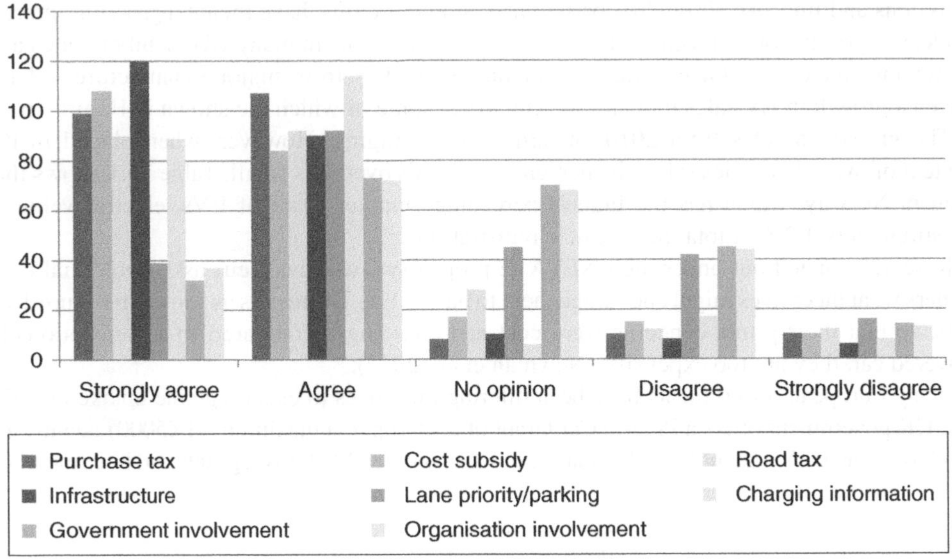

Figure 16.4 Consumer opinions on EV purchase incentives. Source: EVCE/ENEVATE

(Office for Low Emission Vehicles). However, this is available to business as well as private consumers, so this number is likely to contain a number of business and fleet users as well as individual consumers. Similarly, in Norway, there have been a range of incentives put in place including:

- Exemption from purchase tax
- Exemption from VAT
- Access to cities without charge
- Free parking in city centre
- Recharging facilities in city centre with free charging
- Exemption from circulation/road tax
- Use of bus lanes

This has resulted in Norway having arguably the most successful EV market in Europe. However, not all incentives are sustainable in the long term. Also exemptions that allow EVs to enter city centres, such as London or Oslo free of charge, while potentially improving air quality, will not reduce congestion. There is also the argument that whilst governments subsidise vehicles, there is less incentive for the manufacturers to reduce costs. Even with the government grants, EVs are still expensive vehicles to buy, especially when their limited range is taken into consideration, which limits the number of consumers who would be willing or able to purchase one for personal use.

16.3.1 Urban Electric Vehicles

The restricted range of EVs means they are often targeted for use as urban vehicles, where the trips taken are usually short, and the need to reduce noise pollution and air quality is of greater concern. However, there is also often a lack of off-street parking, which makes it difficult for potential EV owners to charge their vehicles, particularly if they live in apartments. This is where car-sharing schemes become attractive for urban use, as it removes the responsibility of finding a suitable charging location as it is provided by the operator. It also mitigates the high cost of purchasing an EV for individual use. Perhaps the most ambitious such scheme currently operating in Europe is Autolib' in Paris and the surrounding Ile de France area (www.autolib.eu). Autolib' is a self-service EV car-sharing scheme, with cars located at stations throughout Paris and the Ile de France. Users can pick up a car from any station and drop it back to any other station, allowing for flexible journeys. The cars are equipped with a touch screen, allowing the users to identify the nearest available station. They also have a help button that connects to the call centre, should the user experience any difficulties. One of the key elements is customer service – booking a car can be done through fully automated kiosks or through smartphone apps, backed up by a call centre, so bookings and subscriptions can be done face to face, which means users can be shown how to operate the recharging and booking systems, removing any anxiety around using a new technology. A further benefit of the scheme is that it is open to drivers from 18 years old and is an attractive proposition for younger drivers, as the cost of owning and running a car is prohibitively expensive for the majority. By summer 2012, there were 1100 stations and parking spaces and 1740 cars. The scheme expanded to Lyon at the end of 2013, with further expansion in 2014, including Indianapolis in the United States. Other smaller schemes have started to appear in cities throughout Europe, such as Zen Car in Brussels (www.zencar.eu) and Mercedes-backed Car2Go with schemes in various cities in Germany, the Netherlands, Austria, the United Kingdom and also in the United States and Canada (www.car2go.com).

Many countries around the world have set targets for EV use, providing further opportunities for such schemes. Thus the United States is aiming to have one million EVs on the road by 2015, and China is aiming for five million by 2020 with annual sales of one million by 2020 and a 20–30% market share by 2030. Similarly, Japan is planning to have two million FCEVs in use by 2025, while South Korea is aiming for 50000 BEVs and PHEVs by 2020 with these constituting 50% of new sales by the 2030s (Shulock et al., 2011).

16.3.2 Rural Electric Vehicles – The Welsh Case

While it is often supposed that EVs are the domain of urban areas, it is important not to accept this received line uncritically. Indeed, an equal, if not more, compelling case can be made for utilizing EVs outside towns and cities. Densely populated areas will tend to display lower car ownership levels generally, alongside well-established public transportation infrastructures and inherently shorter distances to be travelled. As such, for anything other than local air pollution, the potential of EVs to urban sustainability seems rather limited. Of course, toxic emissions are a serious concern and can blight some cities, thus prompting beneficial moves to some form of e-mobility, whether it be in the Autolib model taken on by Paris or in the ground breaking air pollution regulations established in Los Angeles. However, questions of sustainability as they relate to car use are far wider than the end product from exhausts and, in some ways, EVs could contribute to existing problems if promoted in urban areas. Battery EVs will largely be taken on as second cars due to their limited range simply increasing the numbers of vehicles constructed, while promoting them over the equally plausible use of a bus or a bike will have the effect of multiplying the numbers of vehicles on the road. Further, simply encouraging more private vehicle ownership promotes the system of consumer capitalism and, as such, has a detrimental effect on social sustainability – it ensures that the individualistic motif of neo-liberalism continues to hold sway as commuters remain isolated from one another, further eroding notions of community and civic identity. Practically, there are also problems for using EVs in built up areas, not least of which that there is a fundamental lack of (domestic) space within which to park and recharge the vehicles. Moreover, the annual breakeven distance for an EV could easily be in the region of 80 000 km, which would be difficult to achieve for most urban consumers with fixed and limited daily commutes. So, while the contribution of EVs to improving inner city air quality is substantial, as a means of decarbonizing transport it is both expensive and of marginal benefit.

In contrast to urban areas, suburban and rural commuter ranges are typically better suited to the capacity of the EV, with longer ranges (though, in general, within the capabilities of battery technology). Distances travelled for those outside the towns and cities, then, come much closer to fitting the discharge–recharge cycle more effectively. Following this distance, there is an attendant advantage to be offered in terms of operational costs of EVs, which thereby become greater relative to the purchase cost. As a result, the total cost of, at first glance expensive, EV ownership becomes a far more attractive proposition. Remedying some of the practical problems of the cities, suburban and rural households is far more likely to have off-road parking, meaning that domestic recharging could be achieved – which is crucial regarding the piecemeal charging infrastructure and established trends whereby drivers have been seen to be far more likely to charge at home. In addition, these households are far more likely to comprise multiple car users, meaning that an ICE capable of traversing distances longer than the range of the EV is more likely to be located already. In these circumstances, adding EV to the mix is a more valuable addendum as the conventional car realistically needs to be run but there is a genuine prospect of the EV making a difference for shorter trips.

These lesser distances would otherwise need to be conducted by IC car due to essential difficulties over access to services that befall carless households in the countryside. The dispersed nature of residences relative to workplaces, shopping or schools allied to a weaker provision (being increasingly cut, in the United States at least, due to economic recession and austerity) meaning that ordinary life requires access to automobility. Moreover, with less congestion, trip times are more consistent – this reliability is important in ensuring drivers are not left stranded away from recharging points. Most suburban and rural roads (other than motorways) are adequately suited to EVs in terms of the typical road speeds attained. Often ignored, this application needs to be given proper consideration and, indeed, plentiful examples of schemes that use EVs outside urban localities can be found in Europe.

As a particular case study, it is worth considering what is occurring in Wales. Wales is a largely rural nation; most urban activity can be found in the south of the country, particularly in the cities of Cardiff, Newport and Swansea that straddle the nation's sole motorway – the M4. In addition to these cities, there are clusters of urbanization in the once coal-mining valleys, which feature a cramped concentration of towns and interconnected villages, often forming a ribbon-development along the valley. Beyond this, Wales consists of large areas of open space peppered with often isolated and spread out communities, which by conventional thinking would not be conducive to the deployment of EVs. In reality, there are numerous schemes underway and these take three main forms: community car-sharing, tourist opportunities and local authority fleets. The schemes exist because they meet certain local needs, in particular addressing the declining levels of public transport in rural areas, whereby large swathes of the countryside are not served by rail and people often live a significant distance from the increasingly irregular bus routes. Most such households need a car – often two. Fuel is more expensive than in urban areas and this presents a major contributor to rural poverty.

The bulk of car sharing schemes, from Autolib' to Zipcar, tend to focus on urban areas, where their sustainability benefits are generally trumpeted as a means to reduce traffic and air pollution. However, the active and public transport options are most feasible in the cities, yet such provisions are more necessary in the countryside. Because of the isolated location of many villages and the distances that need to be travelled to do almost anything, most households will generally own a car (and many run two or more). In this regard, there are a small number of community car-sharing projects covering isolated communities in some of the more remote parts of Wales. These are to be found in two of three national parks, where Pembrokeshire's Cilgwyn Community Group and St David's Eco City Group are both in the early stages of their operation as is the Brecon Beacons' Cilgwyn Energy Group, while Talybont Energy are relatively well established. Talybont Energy were set up in 2010 and stand out as the most emblematic of these schemes; though they struggled to get going, they are now as strong as ever several years down the line, setting a strong example as to what can be achieved. As with the other three schemes, Talybont Energy have an electric van (which they have named Huelwen; sunshine in Welsh, to reflect the solar panels used to charge it), but they also have a car nicknamed Mr Chips, which has been converted to run on recycled vegetable oil collected from local cafes and pubs. The group has been involved in local sustainability projects for nearly a decade, including a successful hydroelectricity generation scheme. They entered the transport field on calculating that personal transport accounted for 40% of their carbon footprint (Kidd, 2013). The village has around 300 households, and 15 are presently taking part. Charges are intended to be roughly comparable to running a privately owned car for the same journey, though they cover to ensure the vehicles' insurance, maintenance, fuel and electricity costs. However, the ability to hire the vehicles by the journey cedes the additional risks and responsibilities of long-term ownership. It gives access to people that could not otherwise afford to buy and maintain a vehicle personally. Though no users have yet been able to give up running a car, the potential of such a scheme is clear. With an electric van, the club provides a form of sustainable mobility typically considered prohibitively expensive to buy (Newman et al., 2014).

Beyond day-to-day use, picturesque rural areas can also offer abundant tourist opportunities to use EVs in rural areas as visitors will inevitably need to use personal transport to traverse the terrain. Recognizing the value, and the cachet, of being able to offer to do this sustainably, the Brecon Beacon's Eco-Travel Network, in 2012, introduced 6 EVs and 40 charging points (www.ecotravelnetwork.co.uk). The Eco-Travel Network is run as a not-for-profit organization made up of local business reliant upon tourism, such as Bed and Breakfasts, pubs and restaurants. Using a £25k Start-up Grant that they obtained from the Brecon Beacons National Park Sustainable Development Fund, they purchased a fleet of Renault Twizys and established charging facilities attached to their various commercial establishments. Ongoing costs are covered by the private partners who benefit from the tourists brought to the area by the novel application of EVs. The use of EVs is particularly beneficial to these partners because they allow the promotion of the area as an ecotourism destination. In areas that attract tourism, establishing viable niches can be most fruitful and with its beautiful and often barren scenery, and an established environmental reputation as a location for enjoying outdoor activities such as adventure sports and having such low light pollution as to be a recognized dark skies location, it is most appropriate to use EVs to contribute to this image (Newman et al., 2014).

First and foremost, then, these vehicles help to reduce the carbon footprint of visitors – tourists will tend to come by car and then drive long distances during their stay: this is a necessity, many areas are not accessible by public or active transport and commuting between locations would range from difficult to implausible without an automobile. By replacing internal combustion engines with electrics, they can at least claim reduced tailpipe emissions. As such, the project also seeks to promote the environmental benefits of EVs. It is crucial, though, they choose to use Twizys to demonstrate the value of EVs – these quadricycles provide a quite different experience to a car. The Twizys supplant the routine, everyday nature of commuting with something fun, novel and exciting thus offering a genuine holiday experience of doing something different. By highlighting the sheer joy of driving that can be achieved through such EVs, this application provides a most useful advert for the technology appealing to more than just ideology and environmental consciousness; alongside all of their wider benefits, they are enjoyable.

Alongside personal and leisure use, another valuable deployment of EVs is in work fleets, and this can be as applicable in the city as it is the countryside. Indeed, local authorities such as Pembrokeshire, Rhondda Cynon Taff and Wrexham operating largely in rural Wales have been proactive in introducing EVs into their work fleets. As a successful example, Swansea has eight electric cars in their fleet, a large number reflecting the authority being the country's first to appoint a councilor specifically mandated for sustainability. Of course, the council covers Wales' second city but it is also made of the large suburban and rural area that surrounds the urban centre – coast, valleys and farmland. More rural again have been the trailblazing Carmarthenshire County Council, the first to use EVs in Wales, back in 2011. They were able to purchase the vehicles using Welsh Government funding, specifically drawn from the National Travel Plan, which funded five Sustainable Travel Centres in Wales (Newman et al., 2014). As one of these sustainable locations, the government provided Carmarthenshire with £1million, which initially funded two electric cars as a pilot and, in 2012, led to the purchase of two more based on the experience of running the first two. Carmarthenshire is a rural county in South West Wales. It has a population density of 75 km^{-2}, 18th out of Wales' 22 most densely populated counties (National Assembly for Wales, 2008).

The electric cars are used to reduce the significant carbon footprint accrued from commuting across this large, open county. The cars are based in the county town of Carmarthen but they are largely used as outreach for the disparate towns, villages and isolated settlements across the county. As a result, the cars are frequently used in longer, irregular journeys, beyond the average urban commute that it is generally assumed EVs are designed to facilitate. Having kept recorded logs of use and conducted social research into attitudes, the cars have proven generally popular amongst staff – there was initial scepticism and anxiety but, having all undergone compulsory EVs training, these doubts tended to dissipate as staff realized they were a viable alternative to their internal combustion engines. In addition, the cars can be seen to perform a useful secondary function of introducing EV driving to a larger audience such as staff member's families, the people they visit and the public who see the cars in the street (they possess a livery that advertises their being electric).

Examples such as the above can be found throughout Europe, but Wales has an interesting array of schemes amongst a small population of fewer than four million people. So, while EV sales in the region may still be low, these projects show that EVs might infiltrate popular consciousness in new and unexpected ways. The path to transition, then, might not be obvious so it is important to investigate what is going on away from the bright lights of the motor industry and learn from experience at ground level.

16.4 Conclusions

Although older than either petrol or diesel, EVs are nevertheless in many respects a novel technology and technological progress is currently moving at a very fast rate. This means that definitive statements about EVs are premature. Nevertheless from an environmental perspective, as well as a consumer experience perspective, they hold many attractions and are likely to make the breakthrough many now expect as a result of carbon reduction as well as toxic emissions regulation, now featuring in all major jurisdictions (see Chapter 14). In the process, however, issues of electricity feedstocks and demand and distribution are also need to be addressed, as these EVs will place new loads on an existing infrastructure.

Similarly, although long seen as primarily an urban solution – due to its air quality benefits – it is increasingly emerging as a technology equally well suited to suburban and even rural areas. This will greatly increase the applicability of EV technology, providing it with the potential to become a true alternative to the current IC car, at least in the longer term, when some of its current drawbacks have been addressed through a combination of technological advances and changes in customer expectations.

References

Cipcigan L.M., Wells P.N., Nieuwenhuis P.A., Davies H.C., Whitmarsh L.E. and Papadopoulos P., (2012) 'Electricity as a transportation fuel: Bridging the gaps in the electric vehicle value chain', EVVC-2012 European Electric Vehicle Congress, 19–22 November 2012, Brussels, Belgium.

Element Energy (2013), 'Pathways to high penetration of electric vehicles', Final report for The Committee on Climate Change, December 2013, http://www.theccc.org.uk/wp-content/uploads/2013/12/CCC-EV-pathways_FINAL-REPORT_17-12-13-Final.pdf, accessed 12 April 2014.

ENEVATE (2013), 'Final report', http://www.enevate.eu/final_conference2013/report.pdf, accessed 28 March 2015.

Fabio O., Santiangeli A. and Dell'Era A., (2013) 'EVs and HEVs Using Lithium-Ion Batteries', in Gianfranco Pistoia (ed), *Lithium-Ion Batteries: Advances and Applications*, p. 205, Oxford: Elsevier.

Ferdowsi M., Unda I.G., Karfopoulos E., Papadopoulos P., Skarvelis-Kazakos S., Cipcigan L.M., Raab A.F., Dimeas A., Abbasi E., Strunz K., (2010) 'Controls and EV aggregation for virtual power plants', Mobile Energy Resources in Grids of Electricity (MERGE), Eu Grant Agreement: 241399, Wp 1, Task 1.4, Deliverable D1.3, Controls and Ev Aggregation for Virtual Power Plants, 18 October 2010.

Helmers E. and Marx P., (2012) 'Electric cars: Technical characteristics and environmental impacts', *Environmental Science Europe*, **24**, 14.

Høyer K.G., (2008) 'The history of alternative fuels in transportation: The case of electric and hybrid cars', *Utilities Policy*, **16**(2) pp. 63–71.

Kidd A., (2013) *4 Years Experience of a Rural Car Club*, Brecon: The Prospectory.

Morrill J.K., (2008), 'Transportation: Efficiency and Emissions', Chapter 6, *Global Climate Change And Natural Resources: Summaries Of The 2007–2008 Scientific Literature*, (ed. J.E. Morhardt, Claremont: Roberts Environmental Center Press.

National Assembly for Wales (2008) Key statistics for Carmarthenshire, Cardiff: National Assembly for Wales.

Newman D., Donovan C., Nieuwenhuis P., Davies H. and Wells P., 2012 'Market drivers in E-mobility', http://www.enevate.eu/Workpackage3/survey_results.pdf, accessed 19 April 2015.

Newman D., Wells P., Nieuwenhuis P., Davies H. and Donovan C., (2014) 'Urban, sub-urban or rural – where is the best place for electric vehicles?' *International Journal of Automotive Technology and Management*, **14** 3–4, pp. 306–323.

Office for Low Emission Vehicles (2011), 'Making the connection, the plug-in vehicle infrastructure strategy', June 2011, http://assets.dft.gov.uk/publications/making-the-connection-the-plug-in-vehicle-infrastructure-strategy/plug-in-vehicle-infrastructure-strategy.pdf, accessed 12 April 2014.

Office for Low Emission Vehicles (2012) 'Plug in car grant', https://www.gov.uk/government/publications/plug-in-car-grant, accessed 5 November 2013..

Papadopoulos P., Akizu O., Cipcigan L.M., Jenkins N. and Zabala E., (2011) 'Electricity demand with electric cars in 2030: comparing Great Britain and Spain', *Proceedings of the Institution of Mechanical Engineers, Part A: Journal of Power and Energy*, **225** (5), pp. 551–566.

Papadopoulos P., Skarvelis-Kazakos S., Grau I., Cipcigan L.M., Jenkins N., (2012) 'Electric vehicles' impact on British distribution networks', *IET Electrical Systems in Transportation* , **2**(3), pp. 91–102.

Papadopoulos P., Jenkins N., Cipcigan L.M., Grau I., Zabala E., (2013) 'Coordination of the charging of electric vehicles using a multi-agent system', *IEEE Transactions on Smart Grid*, **4**(4), pp. 1–8.

Raab A.F., Ferdowsi M., Karfopoulos E., Grau I., Skarvelis-Kazakos S., Papadopoulos P., Abbasi E., Cipcigan L.M., Jenkins N., Hatziargyriou N. and Strunz K., (2011) 'Virtual power plant control concepts with electric vehicles', 16th International Conference on Intelligent System Applications to Power Systems, September 2011, Hersonissos, Crete, Greece.

Shulock C. and Pike E., Lloyd A. and Rose R., (2011) *Vehicle Electrification Policy Study; Task 4 Report: Complementary Policies*, Washington, DC: The International Council on Clean Transportation.

Tsang F., Pedersen J.S., Wooding S. and Potoglou D., (2012) 'Working paper: Bringing the electric vehicle to the mass market, a review of barriers, facilitators and policy interventions', RAND EUROPE, February 2012, http://www.rand.org/content/dam/rand/pubs/working_papers/2012/RAND_WR775.pdf, accessed 12 April 2014.

Xydas E., Marmaras C., Cipcigan L.M., Hassan A., Jenkins N., (2013a) 'Electric vehicle load forecasting using data mining methods', Proceedings: IET Hybrid and Electric Vehicles Conference, 6–7 November 2013, London, UK.

Xydas E., Marmaras C., Cipcigan L.M., Hassan A., Jenkins N., (2013b) 'Forecasting electric vehicle charging demand using support vector machines', 48th International Universities' Power Engineering Conference, 2–5 September 2013, Dublin, Ireland.

17

Alternatives to the Car

Peter Wells
Centre for Automotive Industry Research, Cardiff Business School, Cardiff University, Cardiff, Wales, UK

17.1 Introduction

The possibilities for personal private mobility obviously extend beyond the car and, to most observers, into objects such as bicycles and motorcycles, three-wheel machines and more. In a great many applications, these other objects are not, however, regarded as substitutes for the car in terms of the functional mobility provided. While other forms of mobility may offer certain distinct advantages (such as lower cost or perhaps greater flexibility), or indeed be regarded as having at least equal performance (a motorcycle can attain the same legal maximum speed as a car), there is a general sense that these other forms of mobility are rather disadvantaged by one or more functional weaknesses compared with the car. The lack of enclosure, the restrictions on the number of people carried or the payload generally, and the lack of versatility compared with a traditional car all render these non-cars as also 'less than cars'.

In addition, the pervasiveness of the car and the sheer scale of the industry that manufactures and supports cars are such that they completely dwarf the other providers of personal private mobility. Indeed it could be argued that cars, in the manner of the cuckoo, have managed to squeeze out other possible road users that were precursors to the car. Moreover, the contemporary car has come to define in legislative, market and consumer expectation terms our assumptions about personal private mobility. The first section of this chapter therefore outlines just how the car has come to be defined in broad terms and how this makes the emergence of alternatives more difficult than would otherwise have been the case. It therefore sets the scene for the playing out of path dependency in automobility as one of the barriers to socio-technical transition towards sustainable mobility.

Despite these observations, there has always been a hidden world of non-car automobility and niche applications where the usual regime dynamics do not apply because of a range of contextual- and technology-specific factors. The second section of this chapter argues that such

The Global Automotive Industry, First Edition. Edited by Paul Nieuwenhuis and Peter Wells.

protected spaces are of growing significance to the wider automotive industry (including commercial vehicles) and are steadily contributing to the battery electric vehicle market – albeit outside the dominant mainstream car market. Thereafter, the chapter gives attention to a relatively neglected phenomenon: the electric two-wheeler (2W-BEV). While still not a substitute for the general purpose traditional car, 2W-BEVs are argued to offer distinct (though possibly temporary) advantages over the car in certain applications and settings to such a degree that there is a transition by stealth underway. By this it is meant that despite the overall neglect of the 2W-BEV by the state in most countries (and the lack of funding support compared with what has been showered upon the mainstream automotive industry), the 2W-BEV has been subject to remarkable technological innovation and achieved substantial market success.

The concluding section argues therefore that the focus on the dominant incumbent factors and the dominant technologies as packaged in a conventional definition of a car runs the risk of missing significant shifts in the structure of mobility provision and the way in which multiple niche markets outside the mainstream may underwrite volume growth in key sub-technologies such as battery packs and control systems that in turn will feed into the mainstream industry.

17.2 Defining the Car: Legislative and Market Boundaries

Cars are defined by legal frameworks in different market jurisdictions. These frameworks are not entirely uniform across different countries, though there has been a degree of convergence over time at a global level. Hence, for example, Japan has long sustained a specific category of small car (the Kei class) that has to fit inside specified dimensions, engine size and performance limits (see Chapter 14). Equally, the Land Rover Discovery is a 'car' in Europe but a 'light truck' in the United States (see also Chapter 15 on such national and local variation). Additionally, tax break points and other features act to shape markets at the national level, despite attempts at the harmonization of international standards with regards to safety and emissions, for example as outlined in Chapter 14. One important exception to the uniformity of markets in the European Union that has relevance to this chapter is that of the French 'voiture sans permit' or what are known more generally as quadricycles. These vehicles are a classic example of something that is less than a normal car; allowed to be used by younger and non-qualified drivers in France because they are extremely limited in terms of power and speed, they are also exempt from the usual safety standards (cf. Nieuwenhuis, 2014, Chapter 7). Traditionally, this market has been served by niche specialists like Ligier and Aixam but offers some interesting possibilities for the application of electric power. Most players have already sought to enter this market, including Aixam with offerings such as the eCity range (which in retail terms starts at €16 999 for the electric version compared with €9999 for the standard diesel). Due to the light weight of the vehicles, Aixam claims a 75 km range and 3.5 h recharge time (see http://www.aixam.com/en/). Unfortunately, the segment as a whole is often considered to be unsafe, and perhaps more importantly most suited to the elderly or those who have lost their 'normal' driving license due to delinquent behaviour.

Another important exception is the 'low volume' type approval process. Known as the EC SSTA (European Commission Small Series Type Approval), it only applies to models built in volumes of less than 1000 per annum (see http://www.dft.gov.uk/vca/vehicletype/ec-small-series-ecssta.asp). The EC SSTA was brought into legal recognition to allow less onerous testing and validation requirements for those manufacturing vehicles in small numbers, and hence to allow some mitigation of the cost of the Type Approval process (see Chapter 14). It has been particularly relevant in the United Kingdom because of the tradition of having many niche car manufacturers who would be unduly burdened by the full Type Approval process. Again this is interesting in terms of the development of niche BEVs, although in other respects, these vehicles are recognized as full cars, not the sub-cars of the type originally found in France.

In the EU, the broad category of passenger cars is termed M1 vehicles and defined under EU regulation 2007/46/EC (as last amended by 385/2009) as motor vehicles with at least four wheels designed and constructed for carriage of passengers, comprising no more than eight seats in addition to the driver.

Moreover, there may be break points in the market that arise out of tradition or local expectations that further act to define the segments of the car market as they are understood by vehicle manufacturers, journalists and indeed the car-buying public (see Table 17.1). Breaking down these ideas is often a difficult task, because it is often hard to understand any logic underpinning them. However, in approximate terms, the major market segments in Europe are given below. Note that the definition of such segments is different in other markets such as North America or Japan and that the categorization of any one model or variant of car into a particular segment is sometimes rather arbitrary, particularly given the contemporary penchant for so-called 'cross-over' models that do not fit neatly into a particular segment category.

A feature of recent changes is that many segments have become subdivided by different quality and price levels according to the power and positioning of the brand. At the lowest entry level, there are the 'value for money' or 'economy' offerings, sitting just below the mass market quality–price compromise brands. Above these two are the traditional mainstream 'upmarket' brands and then a rather rare and specialized 'premium' level. It is notable, however, that in this fairly typical understanding of the market, there is no recognition of the sub-car segment at all, although the A segment in theory could accommodate such types of vehicle. Moreover, the segmentation structure used does demarcate distinct differences in performance and application, but those differences have been somewhat eroded by the versatility of the ICE car in contemporary guise. Hence a sports car used to be rather demanding to drive and maintain and difficult to use in congested streets, but nowadays with advanced electronic gear change systems, power

Table 17.1 The main car segments in Europe

Segment	Broad description	European examples
A	City car, minicars, microcars	Smart For Two
B	Sub-compact car, small car, supermini	VW Polo
C	Small family car	VW Golf
D	Large family car, executive car, midsize car	VW Passat
E	Large car (non-luxury full size)	VW Phaeton
F	Extra large car, full size luxury car	Rolls Royce Phantom
G	Sports	Porsche 911
M	Multi purpose vehicle	VW Sharan
P	Pick up	Nissan Navara
S	Sports utility vehicle	VW Touareg
V	Van	Renault Kangoo

steering and myriad other features, those difficulties have been largely eroded. Similarly, early versions of the SUV (or 4×4) that were explicitly designed for off-road use were not well adapted for road use, generating a lot of road noise, with heavy handling and poor quality in terms of noise, vibration and harshness issues that tend to predominate in the mass market road cars. Yet, often the reviews of such SUVs now remark how the on-road performance of these cars is almost equivalent to mainstream saloon designs.

The point as far as alternatives are concerned is that the understanding of what a car is, and what the car market represents, is defined by these incumbent concepts of segments and the assumed purposes that they are intended to fulfil. The fragmentation of the market over time as discussed in Chapter 3 means that there are many distinct sub-segments or categories of car already filling existing market niches and hence making it difficult to achieve market entry. Moreover, many sub-car alternatives have to 'compete' on the open road with vehicles up to and including 38t articulated trucks. A vehicle like the Sinclair C5 (see http://www.c5owners.com/ for a website for enthusiasts) might have made some sense in itself as a low-cost and simple-battery-powered urban runabout but was utterly unsuited to mixed traffic conditions. In addition, the old pattern of individual car ownership, and the separation of different modes of travel (including the public–private divide), is being broken down, particularly in Europe where there is a strong push towards so-called 'smart' or 'intelligent' transport systems that enable and allow multi-modal and inter-modal transport in an increasingly seamless manner. Intelligent transport systems (increasingly installed as 'apps' on smartphones) have sought to extend beyond the provision of items such as timetable information or the location of an available car in a car-sharing scheme and into the realms of real-time information (for example on congestion on a particular route) and ultimately into ticketing, billing and accounts. In some urban areas (such as Berlin), such systems are relatively well advanced in terms of the uptake and use of multi-modal travel with a wide range of options available (private car, private bike, walking, bus, U-bahn, S-bahn, taxi, shared car, shared bike) and linked to the use of electric power. The better apps are able to inform users of both the range of alternatives for a given trip, and the time and cost involved. In this definition of mobility, the place of the private car is rather more restricted than hitherto, in that there are fewer vehicles in private hands for private use. It is not clear how far such developments in car ownership and use will continue, or indeed how far this multi-modal model can displace car usage.

Nonetheless, one of the main alternatives to the car is to use public transport as and when it is available, and the thrust of many of these developments in travel modality is to seek to reduce some of the perceived negative attributes of public transport to offer a 'seamless' (or perhaps 'painless') travel experience. The densely populated cities of Asia are clearly contenders for the further development of mobility concepts, and this alone may act as a brake on the continued expansion of the car industry paradigm.

17.3 The Hidden World of Non-Car Automobility

Notwithstanding the preceding discussion on the legislative and market boundaries of the existing car market, there is a surprising amount of non-car automobility other than the established and known segments of the minibus, truck, bus and motorcycle. An interesting contemporary example is the quad bike or all-terrain vehicle (ATV). Traditionally, this machine has an internal combustion engine and four wheels, but seats only one person (sometimes two) in the style of a motorbike, with handlebars for steering and no enclosure. While originally designed for and intended for use in rural applications by farmers and those concerned with livestock, the quad bike has become a popular recreational tool and indeed is even used on the road, being captured under current EU quadricycle regulation. Another interesting example is the powered skateboard, often with a tiny on-board two-stroke motor but sometimes with an electric motor and battery, or also the Segway; these devices can be designed for off-road use or on pedestrian walkways. These applications are typical of the emergence of alternative forms of personal private mobility that tend to fall in between existing legislative categories and hence have an uncertain legal status in terms of where they are permitted to be used and by whom. Moreover, many alternative solutions go rather unrecorded precisely because they exist outside the legal frameworks of the day: there is no need to have a license or to register them, because officially there is no recognition that they exist. As a consequence, there is no official data collected on the number in use. Once sufficient numbers of these alternative vehicles do permeate everyday life, there may be an impetus to bring them into the realm of official existence – though that may mean an attempt to ban their use, to control their use, to tax their use or simply to define more precisely when, where and under what circumstances they may be used.

Indeed, the range of potential categories of non-car battery electric vehicles is rather wide: For example, Harrop and Das (2013) identify off-road mining vehicles, buses, light industrial and commercial (e.g. fork lift trucks), recreational and non-road vehicles (e.g. golf carts), mobility aids for the disabled, two-wheel vehicles, military applications, marine applications and a general catch-all of 'other'. It is indicative that, in their view, the total market was of 38.8 million vehicles (of all types) in 2013 and would rise to 116 million vehicles by 2023 but the pure BEV market was just 70 000 in 2013 worldwide, with a forecast to increase to just 2 million by 2023. By 2023, the total world market for electric vehicles would be worth an estimated US$294 billion. Thus, the BEV in the form of the traditional car, which is the one that steals all the limelight, will according to this perspective obtain less than 2% of the total electric vehicle market by volume – though of course the per unit value is relatively high compared with, for example, electric bicycles which are discussed further.

It is an underappreciated feature of the modern era that the single largest emitter of CO_2 in the world is the US military (Karbuz, 2012) and that the financial cost of bringing a volume of petrol or diesel to the front line of military operations is usually a multiple of the purchase cost of the fuel in the first place. The US military has a large fleet of 'non-tactical' vehicles (quite aside from the potential military benefits of deploying electric vehicles in some theatres of operation), which in 2010 amounted to nearly 200 000 vehicles. So for cost reasons, or to reduce carbon emissions, or to reduce strategic vulnerability to supply, the US military has a strong incentive to adopt alternative vehicle fuels and powertrain.

17.4 Transition by Stealth: The 2W-BEV

According to Navigant Research, the global market for 2W-BEVs is expected to reach about 38 million by 2020, with China accounting for 90% of sales (see http://www.navigantresearch.com/newsroom/electric-bicycle-sales-to-reach-nearly-38-million-units-per-year-by-2020). While reliable data is hard to establish in many instances, it is thought that 2W-BEVs number about 180–200 million in China as of 2014, considerably more than the total car population.

What is particularly interesting about the 2W-BEV market is that it emerged more or less spontaneously, without the large supportive R&D programmes evident in the electric car market and without the usual array of incentives and inducements to entice customers into buying and using the products. There seems to be some evidence to suggest that the practice of banning traditional motorcycles (with internal combustion engines) from the major cities of China acted as a stimulus to demand (Yang, 2010), while other research suggests that (the lack of) access to other public transport choices in the rapidly expanding Chinese cities combined with the consequent growing distance from urban centres are also factors (Weinert et al., 2007a, b). The tradition of bicycle use in China (Zhang et al., 2012) certainly provided a platform for the early stages of the industry when existing bicycles could be retrofitted with kits or adapted by the nascent industry. Indeed, even in China, the attitude of government authorities at national, regional and especially city level towards 2W-BEVs has ranged from ambivalent to downright hostile (Fairley, 2005). While the market impetus for their emergence may owe much to the previous banning of traditional motorcycles, and perhaps to the longer tradition of bicycle use in the country, the rapid expansion from a standing start over a period of about 10–12 years (Weinert et al., 2007b) remains a stark testimony to the 'natural' fit of battery technology to bicycles and other two-wheeled machines. In particular, the development pathway has been relatively simple. We may define some distinct categories of electric-powered two-wheel machines intended to carry one or two people:

- Electrically-assisted bicycle – There are two subtypes of electrically-assisted bicycle. The first is a modified traditional bicycle with pedals, chain drive to the rear hub and possibly gears on the rear wheel. It is a human–electric hybrid where the battery and electrical system are imposed on the traditional design. More recently has emerged what is known in Europe as a 'pedelec', which is a purpose-designed electrically-assisted bicycle with a more sophisticated integration of the electrical components, battery and drive system, and with increasingly sophisticated software management systems to integrate the human power input with the electrical power input, typically restricted to 25 km/h maximum.
- Electric power low-speed bicycle – This is also purpose designed with architecture or design frame of bicycles but without pedals at all. They are designed as low-speed 2W-BEVs, with the ancillary components not intended to cope with higher speeds.
- Electric power high-speed bicycle – This type of design has emerged in Europe and North America and is distinguished by having a higher speed than is typical for electric bicycles, up to 45 km/h, equivalent to a traditional moped.
- Electric power scooters and motorcycles – As with the bicycle-derived vehicles, these may be either adapted from traditional designs (that used internal combustion engines) or purpose-designed as electrics. The vast majority to date are of the scooter design, with small wheels and fairing in front of the driver. They are designed for higher speed, easily attaining 50 km/h.

Within the aforementioned categories, there is a growing pattern of micro-segmentation as the market matures, and more specific requirements are met with more specific designs (Muetze and Tan, 2013). Hence, for example, it is possible to distinguish between various versions of low-speed electric bicycle on the basis of where the battery pack is placed (e.g. behind the seat, on the down-tube of the frame supporting the seat) and where the motor is positioned (e.g. front hub, rear hub), among other features.

Unsurprisingly, the emergence of this particular transport mode has not gone uncontested or without attendant problems, most notably with a concern with deaths and injuries arising from the use of these vehicles (Rose, 2012; Yao and Wu, 2012) and with the use of lead–acid batteries in earlier and less-sophisticated designs (Weinert et al., 2008). In a related problem, official statistics may fail to capture the reality of the market. For example, it is only in 2014 that the Netherlands officially defined a category of 'speed pedelecs' that owners were required to register and be qualified for in the manner of a petrol moped. Once such requirements are put in place, and of course data can be collected.

Similarly, 2W-BEVs may fail to emerge in some markets for reasons that are not entirely obvious or reducible to logic. As Walker (2013) notes, the United Kingdom has to date manifestly failed to adopt the 2W-BEV on anything like the scale found in the Netherlands or Germany. In part, this may be due to the lack of appropriate infrastructure, but equally it could be because in the United Kingdom, the bicycle has long ceased to be a mode of transport (in a functional sense) and is rather a recreational toy that is mostly used by people as a means of exercise. In that context, the electrification of the bicycle power train seems like 'cheating' to the users, as Walker (2013) puts it – in the United Kingdom electric bicycles are for people who don't cycle.

17.4.1 3W-BEVs

Electric three-wheel vehicles (3W-BEVs) tend to come in one of the three categories: the sports version, the city runabout and the utilitarian. Many of the recent designs and prototypes are heavily styled with an emphasis on the sports or urban chic characteristics, while of course there has been something of a resurgence of interest in the ICE three wheeler as illustrated by the Morgan 3 Wheeler. The ill-fated US Aptera 3W-BEV was officially defined as a car by the Department of Energy. It was first unveiled in 2007 but the business failed in 2011 having not produced a single vehicle for sale. The project for the electric version (there is also a petrol version) was taken over by a Chinese group called Jonway in 2013, though to date, no serial production has been forthcoming.

The Swiss Cree SAM three-wheel prototype BEV produced in 2001 is typical both of the sports 3W-BEV genre in terms of the product design configuration and the fate of the business. The design had a 'tadpole' layout with two wheels at the front, and one at the back, seating two people in a tandem arrangement. The company produced a number of running prototypes but by 2003 went bankrupt. Despite rumours of interest from a Polish company in the latter 2000s, no further vehicles were produced. Impact Automotive Technologies (IAT, Pruszkow, Poland) do list the vehicle on their website as a 'project' called the SAM Re-Volt but no serial production has been entered into. In 2010, Honda, which has an interest in both the two- and the four-wheel motorized transport market, did show a 3W-BEV that drew heavily on the motorcycle side of the business, though the 3R-C never made it into production. More successful in this context is the Twike, a human power-electric hybrid tricycle originally designed in Switzerland, but now built in Germany. Twikes feature a single front wheel and side-by-side seating for two and have been on the road since the late 1990s and the company appears to be going from strength to strength.

The more functional 3W-BEVs follow the 'tuk tuk' principle with one wheel at the front, handlebar steering, a driver, and then room for two or more passengers behind. An interesting example of this genre is the Swedish-designed Clean Motion Zbee that was explicitly designed to challenge the ubiquitous Bajaj auto rickshaw found across much of Asia. This vehicle was officially unveiled by the Indonesian President Mr. Susilo Bambang Yudhoyono in 2013 and is an interesting package. It weighs only 230 kg and on a full charge can run for 50 km (at 40 Wh/km energy consumption). A somewhat similar concept is the Japanese Terra Motors three-wheel taxi concept which in 2013 was entered into a competition in the Philippines intended to introduce 100 000 EV taxis by 2016 (out of a total of about 3.5 million petrol-powered three-wheel taxis).

17.5 Conclusions

The mainstream vehicle manufacturers have not exactly missed the rise of the 2W-BEV. Since 2012, it is notable that BMW, Audi and Smart, for example, have shown prototypes or even brought vehicles to market in this segment. The BMW pedelec is actually a folding bike that fits into the boot space of the electric i3 model. Meanwhile, the Japanese motorcycle manufacturers Honda and Yamaha are also present in the 2W-BEV market. Traditional bicycle manufacturers have also responded to the emergent opportunities with a growing range of product offerings from industry leaders such as Giant, and from the more expensive specialist producers like Scott. Major automotive component suppliers like Bosch and Schaeffler have also seen a market opportunity to bring much greater sophistication to the emergent 2W-BEV market. Moreover, alongside these established incumbents of the automotive and bicycle industry is a vast array of new entrants all seeking to capitalize on the opportunities for growth that the market appears to offer.

As a consequence, there is a palpable sense of the rapidity of technological change, and of a high degree of market turbulence that is characteristic of a strongly growing and disruptive application, that government and regulators are somewhat struggling to keep up with. The insertion of the wider range (mostly electric) alternatives to the car into contemporary society has been somewhat neglected up to date, perhaps because the applications are seen as rather minor, marginal and unexciting or mere extensions of the existing practice. Yet, the 2W-BEV experience suggests that none of these misconceptions really apply. Moreover, the collective and cumulative impact of all of this growth is to create a viable industry well beyond the confines of the traditional automotive sector and to underwrite a rate of technology development that would not be possible with automotive applications alone.

References

Fairley, P. (2005) China's cyclists take charge: Electric bicycles are selling by the millions despite efforts to ban them, *IEEE Spectrum*, **42**(6), 54–59.

Harrop, P. and Das, R. (2013) Hybrid & pure electric vehicles for land, water & air 2013–2023: Forecasts, technologies, players. Idtechex.com, http://www.idtechex.com/research/reports/hybrid-and-pure-electric-vehicles-for-land-water-and-air-2013-2023-forecasts-technologies-players-000351.asp (accessed 28 March 2015).

Karbuz, S. (2012) How much energy does the US military consume? An update, http://www.dailyenergyreport.com/how-much-energy-does-the-u-s-military-consume-an-update/ (accessed 2 October 2014).

Muetze, A. and Tan, Y.C. (2013) Electric bicycles - A performance evaluation, *IEEE Industry Applications Magazine*, **13**(4), 12–21.

Nieuwenhuis, P. (2014) *Sustainable Automobility; Understanding the Car as a Natural System*, Cheltenham: Edward Elgar.

Rose, G. (2012) E-bikes and urban transportation: Emerging issues and unresolved questions, *Transportation*, **39**(1), 81–96.

Walker, D. (2013) Electric bikes: For people who don't cycle? http://www.bbc.co.uk/news/magazine-21786511, 27 March 2013 (accessed 12 October 2014).

Weinert, J.; Chaktan, M.; Yang, X. and Cherry, C.R. (2007a) Electric two-wheelers in China: Effect on travel behaviour, mode shift, and user safety perception in a medium-sized city, *Transportation Research Record*, **2038**, 62–68.

Weinert, J.; Ma, C. and Cherry, C.R. (2007b) The transition to electric bikes in China: History and key reasons for rapid growth, *Transportation*, **34**, 301–318.

Weinert, J.; Ogden, J.; Sperling, D. and Burke, A. (2008) The future of electric two-wheelers and electric vehicles in China, *Energy Policy*, **36**(7), 2544–2555.

Yang, C.J. (2010) Launching strategy for electric vehicles: Lessons from China and Taiwan, *Technological Forecasting and Social Change*, **77**, 831–834.

Yao, L. and Wu, C. (2012) Traffic safety for electric bike riders in China, *Transportation Research Record*, **2314**, 49–56.

Zhang, H.; Shaheen, S.A. and Chen, X. (2012) Bicycle evolution in China: From the 1900s to present, *International Journal of Sustainable Transportation*, **8**(5), 317–335.

18

New Business Models and the Automotive Industry

Peter Wells
Centre for Automotive Industry Research, Cardiff Business School, Cardiff University, Cardiff, Wales, UK

18.1 Introduction

The themes of business model architecture and innovation have enjoyed increasing attention both in practice and in academia in response to the escalating pace of change in technologies, markets and approaches to business around the world (Schweizer, 2005; Shafer and Smith, 2005; Baden-Fuller and Morgan, 2010; Teece, 2010). As the business landscape has become more volatile and unpredictable, so the certainty attached to previously enduring business models has become eroded. Conversely, business model innovation with and without new technologies is seen as being a key component to competitiveness and also, increasingly, to a new economy of sustainability (Wells, 2013).

Drawing upon diverse disciplines and practice roots, the academic treatment of business models has itself become rather diffuse. Much of the debate has been around two key issues: how is a business model defined and how does change in business models arise? (Chesbrough, 2010; Demil and Lecocq, 2010). In broad terms, a business model can be defined as having three constituent elements: the value proposition that defines how products and/or services are presented to consumers (i.e. how value is captured), the value network that defines how the business is articulated with other businesses and internally (i.e. how value is created) and the context of regulations, incentives, prices, government policy, etc. (i.e. how value is situated within the wider socioeconomic framework). Business models are generally understood to be somewhat dynamic or evolutionary in that they can change over time, particularly with regard to the emergence of a new technology or market. Moreover, a distinction can be drawn between incumbents with an established business model and new entrants or entrepreneurs that might

The Global Automotive Industry, First Edition. Edited by Paul Nieuwenhuis and Peter Wells.

(necessarily) seek to use business model innovation as a means of competition – particularly where it is difficult to compete with established incumbents on the terms of their existing business model.

In the case of the automotive industry, there are both opportunities and barriers to business model innovation for both incumbents and new entrants. While it is usually the case that business model innovation is seen as arising from opportunities presented by shifts in technology or perhaps changing context, rather less explored is the idea that the failure of a dominant business model may itself be the catalyst for change. Equally, business models are generally seen as distinct from strategy, but as is the case in the automotive industry, it may be that the nature of the business model itself allows or constrains certain strategic choices. Moreover, the scope for business model innovation either by incumbents or new entrants may be constrained by issues beyond economic or market logic. That is to say, even a manifestly failing business model may be given wider sociopolitical support if the loss of the business is deemed to be too damaging. At the same time, an innovative business model that better articulates a fit between value capture, value creation and value context may still fail because the prospect of trying to change to the new model is too daunting. Hence, underpinning the notion of business model evolution is that of path dependencies. In an industry as pervasive, large and established as the automotive industry, the question of path dependencies is one of particular significance.

This chapter therefore presents an account of the basis of the contemporary automotive industry business model, followed by an analysis of the pressures for change on this model and the scope for radical and incremental innovation. Much of this discussion is framed by the relative failure, to date, of policies to promote and support electric vehicle sales and use (see Chapter 13). The final section then considers the prospects for future business model innovation in the automotive industry.

18.2 Fundamentals of the Existing Automotive Industry Business Model

At a generic level, the mainstream or high-volume automotive industry has a clearly defined and established business model. Value capture is centred on a product/service package consisting of all-steel vehicle bodies using petrol or diesel engines (with or without hybrid electric drive assistance), which are sold to end-users alongside warranty and service provision. Markets are segmented by various classes and types of car, as discussed in Chapter 2, in a bid to reach different customer types and thereby extract maximum value. Value creation is achieved by the orchestration of in-house capabilities such as the manufacture of vehicle bodies and vehicle assembly, along with procurement regimes that coordinate and manage global supply chains and local clusters (see Chapter 4). Centralized manufacturing operations seeking maximum production economies of scale also necessitate the use of spatially extensive networks of franchised dealers for the sale and support of vehicles, as is outlined in Chapter 9. The value context is perhaps the area where the most change has been evident in recent years, with, in particular, the emergence of large-scale support for electric vehicles by national and local governments in a complex global tapestry of incentives, regulations and other measures. However, as the post-2008 crisis has exposed, governments are also extremely reluctant to forsake the existing automotive industry that is seen as a major contributor to wealth creation, taxation, employment, balance of trade benefits, technological advance and of course personal mobility.

The generic automotive industry business model is founded upon three key innovations arising in the twentieth century: the Ford moving assembly line and related concepts (such as standardized parts, short cycle times), the Budd all-steel vehicle and related concepts (such as painted vehicle bodies) and the Sloan concepts of creating a range of brands and models and related concepts (such as credit for car purchases, used-car trade-ins, annual model cycles) (Marchand, 1991; Raff, 1991; Clarke, 1996; Nieuwenhuis and Wells, 2007). Over time, there has been a process of convergence and consolidation of this business model, as well as of change and refinement. A difficult issue is to determine at what point such changes and refinements amount to an evolution of the business model. The Ford and Budd innovations were the key to defining the value creation aspect of the automotive industry business model, while the Sloan innovations were largely concerned with aspects of value capture. Of course, the two are intimately related. For example, Sloan pioneered the use of paint and colour, along with an emphasis on styling and regular styling changes, as a means of stimulating continuing demand for new cars. This practice was only practicable with the adoption of all-steel body technology, which allowed the use of new paint technologies (Andrews et al., 2006).

Notable among the changes to the business model has been the gradual process, albeit uneven, of vertical disintegration away from the original highly-integrated Ford approach (Langlois and Robertson, 1989). As a consequence, suppliers have a greater share of the ex-works value of the average car than previously and, with time, have taken a greater role in R&D for new technology development. Supplier management has become a central concern for vehicle manufacturers, while global supply chains have become a distinctive feature of the operating environment with suppliers perhaps more able to seek out low-cost manufacturing locations (Cohen and Mallik, 1997).

In terms of consolidation and refinement, there has been a gradual erosion of non-conformist business models (and many businesses) as the Ford-Budd-Sloan model became increasingly prevalent, first in the United States but then in other countries also. It used to be possible to classify the vehicle manufacturers into three main groups: the niche specialists producing low-volume high-performance and luxury cars (Ferrari, Rolls Royce, Bentley, Aston Martin), the middle volume premium producers (BMW, Saab, Audi, Alfa Romeo) and the high-volume mass producers (Ford, Toyota, Nissan, VW, Fiat). These distinctions have increasingly been eroded. Niche producers have often either disappeared or been absorbed into larger groups. The premium producers have sought to retain their 'cost recovery' approach but are increasingly exploring market segments that were previously thought inappropriate and pushing volumes to ever-higher levels. The mass producers have sought to introduce models into 'premium' segments, acquired some niche specialists and/or introduced their own new premium brands. In this framework, important developments such as the Toyota Production System and 'lean' management practices generally are seen as refinements of the established business model, enhancing efficiency and productivity but not actually challenging the fundamentals of value creation, value capture and value context (see Chapter 4). That is to say, in a mature industry with a long-established business model, there is a tendency to organizational symmetry, but the issues of competitive success and profitability arise out of better execution (doing better) rather than doing something very different. Another important refinement, and one that could perhaps underpin a more profound shift in the business model, is that of greatly enhanced manufacturing flexibility arising primarily out of automation and digitization, from R&D through to production and finished vehicle logistics. Chapter 4 outlines some of the key changes in manufacturing in the automotive industry. The resultant flexibility has enabled greater productivity of labour and capital (see Chapters 5 and 6), product diversity, contracted product life-cycles and simultaneously greater 'platform' economies of scale. Through such refinements, the automotive business model has been able to achieve significant increases in product quality and performance at no net increase in real cost.

18.3 Pressures for Change on the Existing Business Model

Despite the enduring success of the prevailing business model, there are also pressures for change arising out of both developments in the value context and from the accumulation of tensions within the structures of value creation and value capture. The pressures within the existing business model derive from the tension between the value creation system and the value capture system – a tension that can also be understood as a contradiction between the demands of the production system and the demands of the market. Notwithstanding the greater production flexibility noted earlier, the manufacturing system remains one in which economies of scale are of primary importance and this, in turn, means that lower per unit costs are achieved by standardized output. Hence the value creation system is one that is oriented towards production ahead of demand, where minimum levels of capacity utilization are needed to achieve break-even volumes. Where markets are expanding rapidly, this value creation system helps to power further expansion by lowering costs, creating a virtuous circle of volume growth and price reductions (or profit increases). The value capture system is aligned with the character of the market, with the focus on selling vehicles and limited service or warranty packages, alongside consumer credit.

However, where markets are not expanding rapidly and where there is a substantial competition from highly similar other manufacturers, market fragmentation may start to erode the mutually supportive structures of value creation and capture.

Chapter 3 outlined how at a global level both characteristics are evident: rapid market expansion in some locations but stagnation and even decline in those countries with the longest experience of automobility. The problems are manifested in phenomena such as overcapacity, rampant discounting and other consumer incentives such as interest-free credit, high rates of depreciation on new cars and a perpetual need to 'massage' cars into the market with tactics such as facelifts, special editions, optional extras and other measures. Overcapacity in particular lies at the heart of the stress on the existing business model and is caused by three features. First, capacity expansion tends to be rather 'lumpy' in that manufacturing economies of scale at a plant level necessitate outputs of say 250 000 units per annum or more, as outlined in Chapter 5 and Nieuwenhuis and Wells (2007). As a result, incremental expansion (and contraction) of output in line with demand is difficult. Second, the constant search for productivity improvements and production efficiencies by every manufacturer tends to result in an increase in available output capacity. Third, there are significant barriers to exit from the industry; the larger the company the more significant those barriers are. Even when a company does fail, it is often the case that the manufacturing assets are transferred to another company rather than removed entirely from the market. The closure of individual plants is a protracted and contentious process, all the more so because their very scale means that the unemployment consequences on localities can be devastating.

A second major source of stress on the business model occurs when there is misalignment with the value context. The very success of the business model in the case of the automotive industry has resulted in the external costs of automobility becoming ever-more apparent with concerns over local issues such as air quality, noise and congestion alongside the pervasive issues of resource consumption, carbon emissions and deaths and injuries from vehicle use. These impacts of automobility were outlined in Chapter 13. Arising from concerns over negative externalities, the governance frameworks that define much of the value context have also changed, albeit in partial and often contradictory ways (Calef and Goble, 2007). As was detailed in Chapter 14, the car as a product has become increasingly regulated and controlled with many implications for the industry. Intriguingly, developments outside the traditional domain of the automotive industry may ultimately prove decisive. Urbanization alongside mobile telecommunications could profoundly change cultures of automobility and the nature of successful business models (Accenture, 2011). Such developments are much more difficult to measure and determine and hence also more difficult for companies to respond to.

18.4 Incremental Business Model Evolution in the Automotive Industry

One important pathway for business model change is that of incremental evolution by incumbent companies. While the dot-com era that stimulated much interest in business model innovation was also somewhat biased towards new-entrant and entrepreneurial business models (Dubosson-Torbay et al., 2002), in a mature and established industry such as automotive, the major incumbent companies have some highly significant advantages that in effect result in barriers to entry for innovative business models. Moreover, incumbents can change their business models, or even run more than one business model, although perhaps in a more constrained manner than a 'clean sheet' new entrant. Incumbents may therefore be restrained by prevailing attitudes and beliefs, by existing skills and capabilities, and by a concern not to jeopardize large business entities that employ thousands of people (Bock et al., 2012). Risk aversion is most acute on core vehicle models, and the evident incrementalism of, say, successive generations of the VW Golf or BMW 3-Series is testimony to the powerful urge not to meddle with a successful formula. Alternatively, incumbent companies have enormous resources, technical capability and market understanding, allowing a protracted period of business model experimentation that may be denied smaller new-entrants and entrepreneurs.

Neither are the vehicle manufacturers strangers to using alternative business models where appropriate. For many years, the practice of using kit assembly plants (known as completely knocked down (CKD), or semi-knocked down (SKD) kits) to serve low-volume markets with high import barriers has been an accepted in industries. Equally, for many years, vehicle manufacturers would subcontract assembly operations to third parties in the event of a lack of their own capacity or, more usually, a concern not to disrupt production lines with derivative models. The willingness to experiment with incremental business models is thus an established facet of the industry, although there is also an evident willingness to revert to the mainstream model as soon as possible. In this regard, the virtual demise of the contract assembly sector in Europe is a direct consequence of enhanced production line flexibility in vehicle manufacturing which has removed the necessity of outsourcing the assembly of niche models.

There has also been some willingness to experiment with incremental business model evolution around the introduction of electric vehicles of various types. The use of novel powertrain features (whether pure battery electric or some form of hybrid) in itself marks a departure from the traditional value capture package. Taken further, pure-battery electric vehicles embody various constraints that have rendered them relatively unattractive to the majority of consumers, but those constraints can be mitigated to a degree by adjusting the business model. Hence manufacturers such as Peugeot have experimented with offering short-term access to a range of other vehicles when a consumer buys an electric model, thereby mitigating the concern that electric vehicles cannot be used for occasional long trips due to recharging problems. Conversely, Renault has sought to offer the Fluence EV separate from the battery pack (which is leased) in an attempt to remove an element of risk from the consumer. It is also interesting to note that Renault with the Twizy has sought to escape the confines of traditional car/motorbike distinctions and design a vehicle with the battery capability in the forefront of their minds. In total, it might be surmised that there has been a gradual evolution in the mainstream business model towards a product-service system approach and away from merely 'shifting the metal'.

18.5 Radical Business Model Innovation in the Automotive Industry

There are many examples of radical business model innovation for the automotive industry, but perilously few successful ones. This alone suggests that the mutually reinforcing elements of the mainstream model remain substantially intact and surprisingly durable, despite widespread expectations that this is an industry that really has to change quite radically in order to be sustainable in the future (Nieuwenhuis and Wells, 1997, 2003; Wells, 2001; Maxton and Wormald, 2005; Wells and Orsato, 2005; Kley et al., 2011). Those attempts that have survived thus far tend to be marginal and small-scale, and hence insufficient to constitute a meaningful threat to the established model. Arguably the most dramatic reinvention of the automotive industry business model was that attempted by Daimler (Mercedes) with the Smart. Almost everything associated with this vehicle was new. It was a new brand, with a novel vehicle concept (two-seat urban car with high fuel economy using a tiny 660cc three-cylinder engine and automatic gearbox), made in a new factory with an unusual cruciform layout, assembled by suppliers working with large modules and sub-assemblies and sold through a novel 'boutique' system of dedicated retail outlets. While not an unmitigated failure, it is reasonable to assert that Smart did not quite achieve what was hoped for by the management. Much of the 'package' offered by Smart pre-figured that now found in the nascent electric vehicle market – for example offering other vehicles on a short-term basis to Smart owners who had a temporary need for a more traditional form of automobility. It is interesting to note that Smart is also a feature of the more recent attempts by Daimler to access the 'mobility services' market in urban areas with their Car2Go concept, with electric versions expected to follow this route (Firnkorn and Müller, 2011). In some regards, Smart may have been ahead of its time, but the multiplicity of novel features of the business model also increased the risks associated with the brand. As an illustration, Daimler had to compensate the suppliers who had invested in the Hambach plant established to assemble Smart cars because the production volumes did not match the initial forecasts. Only by virtue of association with the resources of the Daimler Group was Smart able to continue; as a stand-alone business, it would have failed long ago.

It is perhaps emblematic that one of the most profound redesigns of the automotive industry business model came from an 'outsider' to the industry and ultimately failed, despite much critical acclaim and media interest alongside substantial government support. Better Place sought to commercialize a 'battery swap' business model for electric vehicles, setting up swapping stations to enable vehicle owners to exchange battery packs as quickly as a traditional petrol or diesel 'fill up' (Christensen et al., 2012). Conceptually, the approach had considerable merit offering to assuage range anxiety problems with rapid 'refuelling', alongside the ability to recharge battery packs in off-peak periods as a bulk buyer of electricity in a manner that would optimize grid electricity usage and allow lowest possible costs. Allied to renewable energy sources in countries such as Denmark, Better Place also sought to maximize the carbon reduction potential of electric vehicles, which is heavily dependent upon the source of power used to generate the electricity (Granovskii et al., 2006; Hawkins et al., 2013). The business model failed because not enough manufacturers were prepared to design vehicles to suit the battery swap system (only Renault had made a significant commitment), the vehicles were too expensive and the start-up costs to establish the infrastructure were extremely large. Table 18.1 illustrates briefly some failed business models in the automotive industry.

It is perhaps significant then that the most successful deployment of electric vehicles as of mid-2013 is found in the Paris Autolib scheme (see also Chapter 16 and Nieuwenhuis, 2014, 135–138), which among other features is orchestrated by the local authorities. In a classic instance of market failure, where the provision of cars and infrastructure is mutually dependent, electric vehicles illustrate the continuing significance of direct state intervention to underwrite market development. In this case, the city of Paris organized a competition and awarded the production contract to a relative 'outsider' with the vehicles then assembled at the facilities of Pininfarina, one of the contract assemblers discussed earlier that had increasingly run out of work from the mainstream industry. It is the public sector that also underwrites some of the cost of establishing the infrastructure, and of managing the vehicle fleet, which is then available to visitors and citizens alike on a 'pay per use' or subscription basis.

Table 18.1 Failed innovative business models in the automotive industry

Example	Value creation	Value capture	Value context
MDI Air Car	Franchised factory/retail concept	Zero emissions compressed air engine. Urban vehicle	No specific support for this technology concept
Rydek	Design concept only	Separate body engine vehicle (EV) with batteries; split public/private ownership	Pre-dated support for EVs
Think	Modular assembly in micro-factories. Internet sales plus mobile service support. Later outsourced assembly	2 seat urban EV	Multiple trials and experiments, for example Think@bout London
Better Place	Infrastructure of recharging stations and swap stations; tie in with Renault for initial supply of 100 000 cars. Retain ownership of battery	4-seat Renault Fluence EV. Separate ownership of battery	Sought support in specific locations, for example Denmark, Israel

18.6 Conclusions and Future Prospects for Business Model Innovation

Other 'outsiders' have cast a covetous eye over the automotive sector, and there is a sense that the industry remains ripe for a dramatic reinvention around the themes of mobility, services and information technology (Rishi et al., 2008). There are some intriguing indications that perhaps automobility has 'peaked' in the mature industrial countries and that the personal ownership and use of a car is no longer regarded as an indicator of material success by some people. The utility of car ownership probably has declined somewhat and eroded as it is by the problems of congestion, parking and financial costs through factors such as rising insurance fees.

What is less clear is whether this means the existing business model will reach some form of catastrophic tipping point, or whether the shifts in the value context will be assuaged by the continued incremental development of the mainstream model of value creation and value capture. At present, the most likely scenario is the gradual extension and increasing importance of what are currently rather marginal and insignificant experimental alternative business models. Possibly the development of new business models can occur in parallel with the continued existence of the mainstream business model, though it is not entirely clear how far it is possible for one business to maintain more than one business model, particularly if there are some points of significant conflict.

In essence, there are two options for future business model change in the automotive industry. Thus far, the industry, and particularly the vehicle manufacturers, has remained successful in terms of retaining their primacy in the entire automobility system. The vehicle manufacturers are thus at the centre of the 'industry model', and their own business model is deployed to continue this state of affairs. The main option for the vehicle manufacturers therefore is that they retain control over their own destiny as the 'keystone' component of the entire automobile industry, and any business model redesign is done on their terms. Alternatively, the industry as a whole and vehicle manufacturers in particular may have a new business model imposed upon them, such that they become mere components of a very different 'mobility services ecosystem' in which primary control is vested in another entity. In some respects, this quite profound challenge to the industry is the subtext in the Paris Autolib approach.

In noting this threat, it is also worth noting that the industry has long been highly politicized and that such outcomes may be prevented because of wider considerations beyond the economic utility of one business model over another. It is notable that Better Place sought to establish itself mainly in countries and regions lacking an indigenous automotive industry, for example. However, the logic of the prevailing industry model with the emphasis on production economies of scale is to continue to shift capacity into the rapidly expanding markets of Asia, Brazil and elsewhere while reducing it in the mature industrial economies. Again there is no empirical measure of the point at which, say, VW ceases to be German in most important respects and at which point the political value of the business starts to decline. Faced with long-term trends such as the decline in petroleum supplies and concerns over global climate change arising from CO_2 emissions, the value context that supports the contemporary business model is changing sufficiently to call into question the long-term viability of both the value creation process and the value capture process.

References

Accenture (2011) Changing the game: Plug-in electric vehicle pilots, http://www.accenture.com/SiteCollectionDocuments/PDF/Accenture_Utilities_Study_Changing_the_game.pdf. Accessed 6 September 2011.

Andrews, D., Nieuwenhuis, P. and Ewing, P. (2006) Black and beyond – colour and the mass-produced motor Car, *Optics and Laser Technology*, special issue: Colour and Design in the Natural and Man-made Worlds, **38**(4–6), 377–391, June–September.

Baden-Fuller, C. and Morgan, M.S. (2010) Business models as models, *Long Range Planning*, **43**, 156–171.
Bock, A.J., Opsahl, T., George, G. and Gann, D.M. (2012) The effects of culture and structure on strategic flexibility during business model innovation, *Journal of Management Studies*, **49**(2), 279–305.
Calef, D. and Goble, R. (2007) The allure of technology: How France and California promoted electric and hybrid vehicles to reduce urban air pollution, *Policy Sciences*, **40**, 1–34.
Chesbrough, H. (2010) Business model innovation: Opportunities and barriers, *Long Range Planning*, **43**, 354–363.
Christensen, T.B., Wells, P. and Cipcigan, L. (2012) Innovative business models for sustainable mobility with better place and electric cars in Denmark, *Energy Policy*, **48**, 498–505.
Clarke, S. (1996) Consumers, information, and marketing efficiency at GM, 1921–1940, *Business and Economic History*, **25**(1), 186–196.
Cohen, M.A. and Mallik, S. (1997) Global supply chains: Research and applications, *Production and Operations Management*, **6**(3), 193–210.
Demil, B. and Lecocq, X. (2010) Business model evolution: In search of dynamic consistency, *Long Range Planning*, **43**(2–3), 227–246.
Dubosson-Torbay, M., Osterwalder, A. and Pigneur, Y. (2002) E-business model design, classification, and measurements, *Thunderbird International Business Review*, **44**(1), 5–23.
Firnkorn, J. and Müller, M. (2011) What will be the environmental effects of new free-floating car-sharing systems? The case of car2go in Ulm, *Ecological Economics*, **70**(8), 1519–1528.
Granovskii, M., Dincer, I. and Rosen, M.A. (2006) Economic and environmental comparison of hybrid, electric and hydrogen fuel cell vehicles, *Journal of Power Sources*, **159**, 1186–1193.
Hawkins, T.R., Singh, B., Majeau-Bettez, G. and Strømman, A.H. (2013) Comparative environmental life cycle assessment of conventional and electric vehicles, *Journal of Industrial Ecology*, **17**(1), 53–64.
Kley, F., Lerch, C. and Dallinger, D. (2011) New business models for electric cars – A holistic approach, *Energy Policy*, **39**(6), 3392–3403.
Langlois, R.N. and Robertson, P.L. (1989) Explaining vertical integration: Lessons from the American automobile industry, *Journal of Economic History*, **49**(2), 361–375.
Marchand, R. (1991) The corporation nobody knew: Bruce Barton, Alfred Sloan and the founding of the general motors "Family", *Business History Review*, **65**(4), 825–875.
Maxton, J. and Wormald, J. (2005) *Time for a Model Change*, Cambridge: Cambridge University Press.
Nieuwenhuis, P. (2014) *Sustainable Automobility; Understanding the Car as a Natural System*, Cheltenham: Edward Elgar.
Nieuwenhuis, P. and Wells, P. (1997) *The Death of Motoring?* London: John Wiley & Sons.
Nieuwenhuis, P. and P. Wells (2003) *The Automotive Industry and the Environment; A Technical, Business and Social Future*, Cambridge: Woodhead.
Nieuwenhuis, P. and Wells, P. (2007) The all-steel body as the cornerstone to the foundations of the mass production car industry, *Industrial and Corporate Change*, **16**(2), 183–211.
Raff, D.M.G. (1991) Making cars and making money in the interwar automobile industry: Economies of scale and scope and the manufacturing behind the marketing, *The Business History Review*, **65**(4), 721–753.
Rishi, S., Stanley, B. and Gyimesi, K. (2008) *Automotive 2020: Beyond the Chaos*, Somers, NY: IBM Institute for Business Value.
Schweizer, L. (2005) Concept and evolution of business models, *Journal of General Management*, **31**(2), 37–56.
Shafer, S.M. and Smith, J. (2005) The power of business models, *Business Horizons*, **48**(3), 199–207.
Teece, D.J. (2010) Business models, business strategy and innovation, *Long Range Planning*, **43**, 172–194.
Wells, P. (2001) Micro Factory Retailing: A radical new product, manufacturing and marketing strategy for the automotive industry, in Pham, D.T., Dimov, S.S. and O'Hagen, V. (eds) *Advances in Manufacturing Technology*, London: Professional Engineering Publishing Ltd, pp. 331–336.
Wells, P. (2013) *Business Models for Sustainability*, Cheltenham: Edward Elgar.
Wells, P. and Orsato, R. (2005) Product, process and structure: Redesigning the industrial ecology of the automobile, *The Journal of Industrial Ecology*, **9**(3), 1–16.

19

Future Challenges for Product and Industry

Paul Nieuwenhuis[1] and Peter Wells[2]
[1]*Centre for Automotive Industry Research and Electric Vehicle Centre of Excellence, Cardiff Business School, Cardiff University, Cardiff, Wales, UK*
[2]*Centre for Automotive Industry Research, Cardiff Business School, Cardiff University, Cardiff, Wales, UK*

19.1 Introduction

The past 20 years or so have seen significant advances in both powertrain and body structures. For the latter, we have seen lightweight structures using new materials (other than pressed welded sheet-steel) gradually being introduced, while for the former there have been marked developments in new powertrain solutions as well as alternative fuels. The first volume produced hybrids, the Toyota Prius and Honda Insight were launched in 1997 and 1998 respectively, but now a wide range of hybrids are available. Hybrids have improved, while more advanced versions using a smaller internal combustion engine – dubbed 'range extender' by GM – are now also available, as in the Chevrolet/Holden Volt and Opel/Vauxhall Ampera and the range extended version of the BMW i3. Diesel hybrids have been launched by PSA Peugeot-Citroën, while some fuel cell hybrid cars have been available for lease in limited numbers in Japan and California, with more launched from 2015 onwards. Such technologies sounded far into the future even at the turn of the millennium.

Much of this has been driven by regulation (see Chapter 14), which usually sees the issues surrounding sustainable automobility as primarily, or even solely, technological. Consumers can then be informed and will automatically make the 'right' choice – the largely fictional rational human of classical economic theory is still alive in many corridors of power and influence. The cultural and psychological constraints within which real consumers make their choices have been highlighted in Chapter 4; merely providing consumers with information is not enough to induce more sustainable behaviour. Such traditional positivist policy approaches

The Global Automotive Industry, First Edition. Edited by Paul Nieuwenhuis and Peter Wells.

that provide information and expect consumers to respond with lifestyle changes in a 'rational' manner are increasingly being challenged. Instead people enter such information in their 'discursive consciousness' contesting certain aspects, while accepting others. Inserting more sustainable practices into people's fiercely protected 'lifestyles' becomes the real challenge (Hobson, 2001).

Marketing techniques such as 'nudging' can be used by industry and government to overcome such constraints to promote a greener agenda (Peattie, 1995, 1999). As Hart (1997) reminds us, industry also has a key role to play. Clearly if consumer choice is limited to positive choices, the outcome will be positive. However, where more sustainable practices already enjoy a measure of social acceptance, change is more likely. It is important that a more sustainable car type is still appealing; people should still want it as well as need it – even if such 'needs' are often in the mind of the beholder. For real progress in this area we should therefore look well beyond the technological, yet technology is also our starting point.

19.2 New Engine Technologies

Geels (2005) describes the transition from horse to car, but also highlights the fact that during this phase the car featured a number of competing technologies, notably steam, electric traction and petrol internal combustion. 'Landscape' conditions at that time determined that petrol IC won, especially after the invention in 1912 of the self-starter by Kettering. Steam power was ultimately abandoned due to weight, the potential danger of boiler explosion (largely solved by 1910), but also the inherent loss of efficiency of external combustion, although its advantages were recognized (high torque from standstill, smooth operation, no need for a gearbox, great reliability). The main problem for EVs even then was the range limited by battery capacity. Of course neither steam, nor EVs went away, but became marginalized technologies in the new automotive regime, much like we still ride horses and still use sailing boats. However, with what the 'transitionists' call 'landscape' changes, such marginal species, now confined to niches, can become more viable again and that is what appears to be happening to the EV at present. In ecological terms, we could describe this as a competitive hierarchy with elements of a competitive network (Krebs, 2008, 294; Nieuwenhuis, 2014, 131).

Although, automotive steam power seems to have gone, it is also clear that for the next couple of decades or so we will still have internal combustion (IC) as the dominant technology. Fossil fuel contains a lot of energy for its volume, which makes it difficult to challenge. Fossil fuel and energy that requires fossil fuel in its production (including some current sources of electricity and hydrogen, see below) will get more expensive as it becomes more complex to extract and process, but IC engine efficiency will keep pace for a while. Also, hybrids and plug-in hybrids still use an IC as a range extender in most cases, although that may eventually give way to a fuel cell. But most hydrogen today also derives from fossil fuels, although alternatives such as methanol and ammonia can be derived from renewables. LPG, CNG and biofuels can also be used in IC engines, which makes those attractive, but if we are really going to get in tune with the way things are done on this planet we need to trap more directly into the fresh solar energy that is thrown at us on a daily basis. We need to move away from finite fossilized solar energy and run our EVs on direct solar energy through photovoltaics, wind power, etc. In some respects, the electrification of the car is an inevitable trajectory, while Huber and Mills (2005) argue that a shift towards more electric energy use is inevitable in a broader sense and in line with long-term energy trends going back millennia. They point out that energy is effectively unlimited, rather it is our ability to harness it that is limited. Yet some of these technologies require new infrastructures, which can provide substantial barriers to such a transition.

Volume produced EVs capable of using such sustainable electricity became available around 2010, with Mitsubishi, Nissan and Renault among the mainstream carmakers leading this trend. This also lent more credibility to specialist EV producers, such as Tesla who are positioning their EVs as premium products, design-led in the way Apple products are perceived as premium products. Mainstream producers are now challenging Tesla in the luxury EV segments they have established with products such as the range extended BMW i8 and GM's Cadillac ELR, while Fisker's Karma is due to be revived by Chinese investors in California.

In reality, a direct comparison of EVs with IC cars is unrealistic – they are a different kind of product. Modern EVs contain a lot of novel technology and esoteric materials; while modern hybrids combine state of the art IC technology with state of the art EV technology, so they are bound to cost more than a conventional IC car – why would we even expect them to be directly competitive? Tesla's argument is therefore sound – pay a premium for an EV in the way you pay a premium for a Macbook Air. This does of course require a comparable functionality and this is a more challenging case to communicate to potential customers and investors. Although EVs have a different functionality from IC cars, the functionality comparison depends on how you value range against such features as ease of operation, silence in operation, ability to 'fill up' at home, lower running costs in terms of energy, maintenance and repair, smoother acceleration, etc. From an environmental perspective, only EVs give the option – provided a zero carbon generating mix is developed – of ultimately carbon-free transport.

One model of future, more sustainable motorized personal transport was proposed by Mitchell et al. (2011). They argue that most growth in car demand this century will take place in the Far East and that the wealthier areas here are also increasingly urbanizing; an urban vehicle is therefore the most suitable solution to provide the flexibility public transport cannot manage, even in densely populated areas. In urban environments, even very extensive ones, average daily mileage for a car is limited making electric vehicles particularly suitable. In addition, the need to reduce toxic emissions in more densely populated and motorized urban environments favours zero emission EVs. The small two-seater vehicles proposed by Mitchell et al. (2011) have autonomous driving and are fully integrated, or 'connected', with their environment, aware of other vehicles and obstacles, thus avoiding accidents and as a result the need for heavy crash structures and airbags is also avoided helping reduce weight still further. To what extent this is still 'driving' is open to debate, of course. Perhaps a model could emerge whereby cars are controlled by the infrastructure in urban areas, but by the driver outside the cities, with the possibility of intervention from external systems in extreme situations to avoid accidents. It is clear, however, that the automotive industry is currently firmly wedded to this trajectory, even though many in the industry recognize that these final ends may not be reached. The technologies developed along the way – it is argued – will provide wider benefits.

It is also important to realize that laudable though these new, often cleaner, technologies are none of this will make the car sustainable if we continue to produce these sophisticated products at the rate of over 70 million a year. Much of the move towards cleaner, greener and safer cars is driven – albeit under some duress from legislators – by the mainstream incumbent car companies, efforts from the likes of Tesla and Fisker notwithstanding. These moves towards connected electric mobility mean that car manufacturers now have to talk to electricity providers, infrastructure providers and they need to develop new business models, for value is added differently along the newly emerging EV and connected car value chain. All this means significant change, which is why some new entrants can succeed in an industry traditionally featuring very high barriers to entry. Ultimately we could see a completely different kind of automotive industry emerge from these ongoing trends in both powertrain and structure technologies, as well as changes in use – resulting in part from such changes – highlighting the socio-technical nature of these major transformations. Yet the debate at present is still largely technological. In this book we have tried to address this issue, by discussing the technological alongside social, psychological, organizational, regulatory and other human factors.

The regulatory drivers should not be underestimated; a number of jurisdictions have introduced specific targets for battery electric and plug-in hybrid vehicles (PHEV), for example as outlined in Chapters 14 and 16. However, the benefits of EVs in reducing GHG emissions in areas with high carbon generating capacity – such as China – is limited, although electric powertrains have a number of inherent environmental and efficiency advantages over proposed alternatives such as biofuels, CNG or LPG. Over time, liquid fuels will also become higher in carbon content due to increasing use of CTL (coal-to-liquid), tar sands, etc.; while electricity will become lower in carbon as a result of the growth in renewable and nuclear power. Kendall therefore argues we should reduce our dependence on liquid fuels in any case (Kendall, 2008). This will increasingly serve to support the EV in favour of IC, as the need to reduce carbon from our economies becomes a more pressing policy priority.

The challenge is how to introduce EVs to the public and also how fast. EVs are still often seen as primarily an urban solution, however research at the Electric Vehicle Centre of Excellence (EVCE) at Cardiff University has shown that EVs can be equally useful in rural settings, as outlined in Chapter 16. Several rural EV initiatives are taking place from Bavaria, California and Denmark to Wales to underline this point. EVs can therefore find application much as current IC cars. Of course in the light of slow replacement demand, it will require many years to transform the stock of cars in circulation to EV and reap the full benefits in terms of emissions. It is also worth pointing out that EVs currently use several materials and technologies that may themselves be in short supply within a generation – the search for alternatives is therefore crucial.

Management consultancy McKinsey assessed a range of carbon reduction options for the global vehicle fleet, yet ultimately favoured EVs (McKinsey, 2009). The McKinsey researchers argued that carbon emissions from IC can be significantly reduced, but ultimately the scope is limited. Based on their calculations and estimates, incremental battery electric vehicle costs would decline from €36 000 in 2009 to €5 800 by 2030 for a range of 160 km. Such figures should be seen in the context of typical cost reductions in the mass production automotive industry. An earlier McKinsey study (McKinsey, 2003) found that the car industry typically reduces costs per car by around €3000 every 13 years (= roughly two model generations). At the same time, manufacturers typically add around €4000 of content to vehicles. This leads to a gradual increase in vehicle costs, but for Europe at least, the overall picture for all three electric technology options turns out better than for the other major automotive regions, because of high fuel costs. Thus both HEV and PHEV provide the owner with a net benefit over the first five years of ownership compared to IC while, for the same reason, the additional cost of a BEV is lower in Europe than in North America, Japan, or China (McKinsey, 2009, 20).

19.3 Owning or Sharing?

It is not only the car itself and the way it is produced that need to evolve, the way we use it needs to be subject to a similar evolution as part of new business models (e.g. Wells, 2010, 2013). Private ownership of cars, as well as private or personal usership may also have to change. While some of us may still own cars in future, many of us won't, but will still have the use of them. This is where car sharing schemes, or car clubs are so important a development. The Autolib scheme in Paris, discussed in Chapters 16 and 18 is a good illustration.

One interesting aspect of the emergent multi-technology future for the industry is that each particular package of technology within a broad portfolio of possible solutions will have the most appropriate use applications. In simple terms, for example current industry thinking is that the EV might be best suited to urban use (though see the discussion on rural EVs in Chapter 16), while range-extended hybrids might suit combined urban and suburban use, and fuel cell vehicles will be best for inter-urban long-range travel. This means that the EV optimized for short-range urban commuting and local trips will be less functional for multiple occupants on a long trip. Perhaps the industry solution to this problem is that households will individually own multiple cars, and select the most suitable to the task in hand. While the number of multiple car households has increased over the years, there are clear limits to the ability and willingness of households to embrace multiple vehicle ownership.

Hence, the logic behind the appeal of various forms of collective or shared ownership or access, of which there are many possible variants, is that to achieve environmental and economic optimization of the various possible technologies it will be necessary to match the technology package more closely to the specific trip under consideration. In turn, the fragmentation of the market in terms of technology types and product design variants will further be reinforced by fragmentation in terms of ownership and use variants, and these combined developments are likely to put further pressure on the prevailing industry business model.

19.4 The Future Car

One thing that seems obvious is that making more than 70 million cars a year worldwide is environmentally unsustainable. This will be even more the case with increasing electrification of the powertrain, as the embedded carbon then becomes greater than the carbon emitted in use (Ricardo, 2011; Hawkins et al., 2012). So, the cars we do make, whatever they are like, will have to be made with lower carbon content, which is most easily achieved by making them in smaller numbers and by making them to last significantly longer (Nieuwenhuis, 2014).

However, as explored in Nieuwenhuis (2014), cars then also need to grow and develop with their owners in some way – something that may be best achieved through modularity. Modularity was a feature of the car in its early days where chassis and body were often sourced from different suppliers direct by the customer and such bodies could be changed over time – cars could relatively easily be 're-bodied' and were in many cases. Engines and transmissions are still to a large extent modular and can still be upgraded relatively easily. New developments such as 3D printing also open up avenues for customization either by the owner, or on behalf of the owner by local specialists.

That the car should also be smaller and lighter is clear, in line with Amory Lovins' Hypercar concept, although it will still need to be tough enough to last several decades and capable of development. This may require different ways of making cars. In this context, a few cars and car concepts may be clear pointers to such future vehicles, notably cars like the GMD T25 and T27, Riversimple, Axon, but also Tesla, Nissan Leaf, Renault Zoe and Twizy, Smart, the original Insight from Honda, and the BMW i3 and i8. The latter are significant in that they are not ‘just’ EVs, but also introduce a new way of making cars. In a sense BMW reintroduces modularity in that these cars have a separate carbonfibre body structure fitted to an aluminium chassis. This gives the potential to mix and match different body and chassis units, as previously proposed by GM’s AUTOnomy and its ‘skateboard’ concept. Such developments will clearly also have implications for skill levels and the organization of labour, whereby higher skill levels and a greater range of skills, a process initiated under ‘lean’ production will become even more important.

In addition, cars will become ICT platforms for which suppliers develop ‘apps’. Each customer/owner would choose apps useful for his or her type of car use, while other apps would not be used. Future cars will then be much lighter, at least in structure, more durable, more reconfigurable and adaptable, more integrated with their infrastructure and in many cases shared. They are also most likely driven by electricity, either directly stored on board, or generated on board by some kind of range extender. This range extender can be a fuel cell, or an IC engine, or some other solution. The industry has been somewhat surprised by the announcement, in 2014, by Google that it intended to bring to market autonomous cars of its own design. This particular development not only suggests a radical restructuring of the industry, it also hints at a very different culture of automobility that might arise out of autonomous cars. Mainstream manufacturers have sought to link the autonomous car concept with that of safety, arguing that only through such active systems can road traffic deaths and injuries be reduced significantly. On the other hand, the results in terms of the emotional engagement with automobility are uncertain: is this the comfort of cocooning or the imprisonment of humanity into little pods over which there is no direct control?

Finally, we must recognize that the future car may in many instances not be a car at all (Wells and Lin, 2013; see also Chapter 17). The industry has only started, in 2014, to appreciate the significance of the electric two-wheel and three-wheel revolution that is mostly underway in Asia. These vehicles that maximize the power to weight benefits of electric traction are comparatively cheap to buy and run, fit into congested urban settings, and suit the daily travel requirements of many millions of people. Sales of electric two wheelers have boomed in countries such as China without the support of the state for R&D, without the participation of the mainstream automotive industry, without incentives or inducements to consumers, and often in the face of outright regulatory hostility by city authorities keen to exclude such devices from their roads. None of this has mattered, which speaks volumes for the fundamental appeal of the technology in this application. A key challenge for the future will therefore be to develop the regulatory and infrastructure space within which real low-carbon mobility can develop. Heavy- and fast-electric cars are a rather ungainly application in terms of material and energy efficiency.

19.5 The Future Industry

It is striking how many of the chapters in this book deal with the struggles of this industry to maintain some degree of stability and cohesion in a turbulent world. As was evident in the cases of labour-management relations in North America and Japan, in many regards it is labour that has borne the brunt of the economic pressures brought to bear on the industry. Equally, as was clear in the cases of India and South Korea, those more recent entrants to the global automotive industry have not been immune to this corporate and economic turbulence. Markets are inherently chaotic, and businesses seek to try to impose some degree of coherence onto markets in order to provide the conditions for successful accumulation of profits, but at least in the case of the car industry it seems that the imposition of order onto a disorderly world is becoming increasingly challenging. Not least, the potential sources of disruption are burgeoning as the industry extends its spatial reach and migrates into other technological competences.

Moreover, across the whole value chain there is likely to be a shift in the balance of value creation and employment towards process that are downstream of original manufacturing. It is not clear at present what these developments will mean in terms of employment, skills and working practices. For example, if vehicle manufacturers retain ownership of the vehicle and derive an increasing proportion of their output from remanufacturing, refurbishment and updating products that are returned from the market then it is likely that this sort of work will be more difficult to automate and to disintegrate into narrow tasks. Perhaps workers will be required to be more polyvalent, and prepared to shift from the assembly process to operations traditionally associated with dealerships (such as maintenance and repair of vehicles).

Indeed, in some respects it could be argued that all businesses will be new entrants in the emergent new automobility regime of the future. To be sure, some of these new entrants are likely to be from the ranks of the existing car industry. Perhaps in this new industry the actual manufacture of cars will be less important, and car manufacturers may become product suppliers to other businesses that are mobility managers (and perhaps Autolib is an interesting template for this model). Perhaps the existing industry of the old 'west' will become isolated into ever-smaller upmarket niches as the low-cost locations of Asia begin to erode market share in the more commodity-like segments. Perhaps the industry will indeed finally consolidate into five or six major companies running portfolios of brands. Perhaps even the emergent new technologies will usher in a repeat of the early years of the industry, when many new businesses sprang into life around the widespread availability of core 'building block' technologies (see Chapter 5). Many of these technologies are viable at low volumes, and of course the capabilities of three dimensional printers are improving apace. So new product concepts serving localized markets with highly customized designs might yet be a feature of the industry in the future.

In addition, the forward focus of many analysts and researchers, while inevitable, tends to neglect the question of what happens to the existing producers, the existing products, and the existing user practices. This question of 'persistence' as it is known in the transitions literature is starting to attract attention, in part because of the growing recognition that such persistence is of substantive importance. It is notable that there are many enthusiasts with respect to cars, people who seek to retain in use classic products from the past – supported by companies making available replacement parts. Little attention has been directed at how quickly the existing industry as has been characterized in this book will recede. Will dealer networks become a retail casualty in the manner of, say, video rental stores or (in the United Kingdom) pubs? Will existing fuel stations supplying petrol and diesel expand into alternative fuels, or decline into marginality as their market is eroded? In the United States there is already much talk of the demise of the classic shopping mall as the demographic, cultural, retailing, and economic changes sweeping through the country render these suburban palaces of consumption increasingly outdated. Will traditional logistics and 'freight forwarding' companies be rendered redundant as production shifts to become near to the market, late configuration and modular in supply? Will the focus of economies of scale become centred on the brand rather than manufacturing per se? Powerful brands remain a core barrier to entry for new businesses casting covetous eyes on the car market, but there are also plenty of powerful brands outside the existing industry that may be able to migrate that brand into this market.

How will regulators view the cars of the future? The contemporary monocular focus on carbon emissions is understandable, but equally fails to capture the full range of harmful impacts from car production and use. In the future other metrics may be needed. Rather than g/km of CO_2 perhaps a measure of efficiency should be kWh/km. Moreover, there is much to be said for a lifecycle analysis approach, particularly when there will be a reduction in the relative proportion of in-use environmental burdens and an increase in production environmental burdens. While the logic is clear, the practical implementation of life cycle analysis into regulation is hugely difficult given the data needs and the problems of reconciling different variables. In addition, the concern to reduce road traffic deaths and injuries is leading to active safety systems that in various ways will act to reduce or constrain direct driver control over the vehicle – but with potential consequences in terms of liability, and also for issues such as data protection and personal privacy. Perhaps as the scientific base of knowledge on climate change and other environmental impacts improves, and as the reality of the lack of current progress on such issues becomes ever more apparent, so the need for more stringent and less consensus-based regulation may become evident, resulting in more confrontational positions between regulators and the industry.

How quickly will cultures of automobility change? It is fascinating to note the continued presence of subcultures from the past in contemporary society, such as Hells Angels on large motorbikes or (in the United Kingdom) 'mods' on Italian scooters. Cars can be enduring as products, given enough care and attention, even if their functional utility may decline with time. The aspirational nature of cars, even though subject to some challenge, does not yet appear to have been entirely displaced. Perhaps in the future the ownership and use of a profligate car (i.e. one that is unreasonably heavy, large, resource intensive, fuel consuming, etc.) will come to be regarded as the equivalent of smoking in public, a fundamentally anti-social activity for which people should be castigated rather than admired. In a world increasingly characterized by a bifurcation of wealth and opportunity, how far will opulent lifestyles be admired or denigrated?

If nothing else, this book has sought to highlight that from our understanding of the past and present of this industry, the future is more uncertain than ever before. As a consequence the traditional forecasting tools of the industry will become of declining relevance to inform strategy or indeed the shape of future regulation. In this more complex world there will be fewer simple answers and, probably, more failures. Out of this maelstrom a new industry will emerge, but one unlikely to be characterized by global uniformity.

References

Geels, F.W. (2005), 'The dynamics of transitions in socio-technical systems: A multi-level analysis of the transition pathway from horse-drawn carriages to automobiles (1860–1930)', *Technology Analysis & Strategic Management*, **17**(4), 445–476.

Hart, S. (1997), 'Beyond greening: Strategies for a sustainable world', *Harvard Business Review*, **75**(1) Jan–Feb, 66–76.

Hawkins, T., B. Singh, G. Majeau-Bettez and A. Hammer Strømman (2012), 'Comparative environmental life cycle assessment of conventional and electric vehicles', *Journal of Industrial Ecology*, **17**(1), 53–64.
Hobson, K. (2001), 'Sustainable lifestyles: rethinking barriers and behaviour', in J. Murphy and M. Cohen (eds), *Exploring Sustainable Consumption; Environmental Policy and the Social Sciences*, Oxford: Elsevier, 191–209.
Huber, P. and M. Mills (2005), *The Bottomless Well; The Twilight of Fuel, the Virtue of Waste and Why We Will Never run Out of Energy*, New York: Basic Books.
Kendall, G. (2008), *Plugged In; The End of the Oil Age*, Brussels: WWF.
Krebs, C. (2008), *The Ecological World View*, Collingwood, VIC: CSIRO Publishing.
McKinsey (2003), *HAWK 2015: Knowledge-based Changes in the Automotive Value Chain*, New York: McKinsey & Co.
McKinsey (2009), *Roads Toward a Low-Carbon Future: Reducing CO_2 Emissions from Passenger Vehicles in the Global Road Transportation System*, New York: McKinsey & Co. Inc.
Mitchell, W.J., Borroni-Bird, C.E. and Burns, L. (2011) *Reinventing the Automobile: Personal Urban Mobility for the 21st Century*, Boston: MIT Press.
Nieuwenhuis, P. (2014), *Sustainable Automobility; Understanding the Car as a Natural System*, Cheltenham: Edward Elgar.
Peattie, K. (1995), *Environmental Marketing Management: Meeting the Green Challenge*, London: Prentice Hall.
Peattie, K. (1999), 'Trappings versus substance in the greening of marketing planning', *Journal of Strategic Marketing*, **7**, 131–148.
Ricardo/Carbon Trust (2011), *International Carbon Flows – Automotive*, London: Carbon Trust; www.carbontrust.com/media/38401/ctc792-international-carbon-flows-automotive.pdf (accessed 13 April 2014).
Wells, P. (2010), *The Automotive Industry in an Era of Eco-Austerity; Creating an Industry as if the Planet Mattered*, Cheltenham/Northampton: Edward Elgar.
Wells, P. (2013), *Business Models for Sustainability*, Cheltenham/Northampton: Edward Elgar.
Wells, P. and X. Lin (2013), Spontaneous emergence versus technology management in sustainable mobility transitions: Electric bicycles in China, Paper prepared for the Sustainable Consumption Research and Action Initiative (SCORAI) conference, Clark University in Worcester, Massachusetts, USA, 12th to 14th June.

Index

The Global Automotive Industry, First Edition. Edited by Paul Nieuwenhuis and Peter Wells.

Printed and bound by CPI Group (UK) Ltd, Croydon, CR0 4YY
27/10/2024
14580361-0002